Chemical Demonstrations

Chemical Demonstrations
A Sourcebook for Teachers
Volume 1, Second Edition

Lee R. Summerlin
The University of Alabama at Birmingham

James L. Ealy, Jr.
The Peddie School

AMERICAN CHEMICAL SOCIETY

WASHINGTON, DC   1988

Library of Congress Cataloging-in-Publication Data

Summerlin, Lee R.
  Chemical demonstrations: a sourcebook for
teachers; Lee R. Summerlin, James L. Ealy, Jr.;
2nd ed.

  p.   cm.

  Includes index.

  1. Chemistry—Experiments.

  I. Ealy, James L., Jr. (James Lee), 1943-   .
II. Title.

QD43.S77   1988   542—dc19   88-21009

ISBN 0-8412-1481-6 (v. 1)

ISBN 0-8412-1535-9 (v. 2)

About the Authors

LEE R. SUMMERLIN received his B.A. from Samford University, his M.S. from Birmingham–Southern College, and his Ph.D. from the University of Maryland. He has held many teaching and administrative positions and has served as a consultant to various organizations, companies, colleges, and school boards in this country and abroad. He has served as a peer review panelist for several science education development programs and as a chemistry consultant to various National Science Foundation Institutes. He has presented many seminars and published a number of books on methods and aspects of teaching chemistry and science. He has also conducted numerous workshops on chemical demonstrations throughout the country. He was codirector of the Institute for Chemical Education program at the University of California—Berkeley in 1985 and coordinator for the program in 1986, 1987, and 1988. Summerlin has held office or had major committee assignments in the National Science Teachers Association, the American Association for the Advancement of Science, and the American Chemical Society. He is a member of the ACS Task Force on High School Chemistry. He received the James B. Conant Award (1969), the Florida Section Outstanding Chemistry Teacher Award (1967), the Gregg Ingalls Outstanding Teaching Award (1985), and the Chemical Manufacturers Association National Catalyst Award (1986). He is coauthor of *Chemical Activities,* published by the American Chemical Society.

JAMES L. EALY, JR., received his B.A. in chemistry and biology from Shippensburg University and his M.Ed. from Lehigh University. Since that time, he has taught at the Mercersburg Academy, the Hill School, and St. John's School, and is currently teaching at the Peddie School in New Jersey. In each of these schools, he worked to develop demonstrations and courses that would encourage students to learn to use the scientific method as well as to instill enthusiasm for chemistry in his students. Ealy has offered advanced placement (AP) workshops for more than 12 years to hundreds of teachers from the United States and several foreign countries. He recently served on the American Version Rewrite committee for *ALCHEM,* a popular high school chemistry textbook in Canada. Ealy has presented various high school chemistry, AP chemistry, and chemical demonstration workshops at the Biennial Conferences on Chemical Education and at several CHEMED conferences. He is an advanced placement reader and consultant for the College Board. He has served as a subject editor for the Secondary School Section for the *Journal of Chemical Education* and is serving his fifth term on the American Chemical Society Advanced Test Committee II. Ealy recently designed the AP chemistry program for the Johns Hopkins Center for Gifted and Talented Youth and currently teaches that program at the Franklin and Marshall College campus.

Contents

Acids and Bases

Energy Changes

Equilibrium

Kinetics

Oxidation–Reduction

Colloids

Polymers

Appendixes

Preface

This book contains more than 100 demonstrations suitable for use with an introductory chemistry program. These demonstrations are simple, effective, and enjoyable. They can be used to introduce many major concepts in chemistry.

Our purpose in presenting these demonstrations is not only to provide the chemistry teacher with a sourcebook of ideas but also to promote chemical demonstrations as a teaching technique to be used with the blackboard, textbook, and laboratory. Performing demonstrations in class is an effective way to make chemistry more understandable and can be fun for teachers and students.

This second edition offers several changes in compliance with new policies established by the ACS Books Department. One such policy has to do with the use of known or suspected carcinogenic chemicals. We have not included in this volume any demonstrations that use substances classified as carcinogens by the National Toxicology Program. We have also excluded demonstrations that involve an unacceptable risk to teachers or students. Unfortunately, this means that some of our long-time favorite demonstrations must be excluded. However, we firmly believe that the well-being of teachers and students should never be compromised, and we applaud ACS Books for taking this stand and setting an example that we hope other authors and publishers will follow.

These demonstrations are presented in a simple format for quick reference: We tell you what the demonstration shows, how to do it, what the reactions are, and how to prepare the solutions. We give special safety instructions and teaching tips, including background notes and questions suitable for use with the demonstrations. We encourage you to use this text as a sourcebook: Add notes and demonstrations that work well for you in your classroom or laboratory.

We suggest that the teacher keep the following in mind regarding demonstrations:

1. Do not think of demonstrations as a replacement for the laboratory. Nothing can substitute for hands-on experience provided by doing laboratory experiments. Rather, demonstrations should be thought of as an extension of the laboratory, another opportunity for students to become more astute observers.

2. Demonstrations should actively involve students. Chemistry is not a spectator sport. Although the teacher performs the demonstrations, students should be involved as assistants whenever possible. Remember, though, that students serving as assistants must observe the same precautions as the teacher.

3. Demonstrations should be simple and easy to understand. Quite often we overlook effective demonstrations because they seem so simple. The demonstrations in this book do not require exotic chemicals or elaborate equipment; nor do they introduce concepts outside the scope of the general chemistry program.

4. Demonstrations should catch and hold student interest. They should be short and catchy. To achieve these goals, the demonstrations in this book involve color changes, gas evolution, colorful precipitate formation, and other obvious chemical changes. They are intentionally designed to be enjoyed by students.

5. Demonstrations should work. Always rehearse them before class. Many factors determine whether a demonstration will work or not. The age of chemicals and the concentration of solutions can make a big difference. Even if a demonstration has been performed dozens of times, check it to ensure that everything works. Remember, to experiment means to try, but to demonstrate means to show.

Properly acknowledging the originator of most demonstrations is difficult because chemical demonstrations have been around for a long time and have undergone many modifications, including our own. However, we gratefully acknowledge all efforts of chemists, past and present, who have tried to make chemistry more interesting and understandable by developing demonstrations. We also express our appreciation to the hundreds of chemistry teachers and students around the country who share with us the philosophy of sharing ideas and who have used our demonstrations and offered suggestions for improving them.

We are especially indebted to several individuals: Christie Borgford and Julie Ealy, coauthors of *Chemical Demonstrations, Volume 2,* have allowed their wealth of ideas to flow over into this new edition. Bruce W. Brown of Portland State University has worked with us on safety aspects of the demonstrations. We appreciate the many hours spent with the Ad Hoc Task Force To Advise on Safety Considerations in ACS Books, and the guidance provided by its members: chair, Gary Long, James A. Kaufman, W. H. Norton, Douglas Walters, and Jay A. Young. Of course, the outstanding staff of the ACS Books Department, Joan Comstock, Janet Dodd, and our editor, Paula Bérard, make writing a book an enjoyable and exciting event.

Lee R. Summerlin
University of Alabama at Birmingham
Birmingham, AL 35294

James L. Ealy, Jr.
The Peddie School
Hightstown, NJ 08520

Properties of Atoms

1.   Electronegativity, Atomic Diameter, and Ionization Energy

A model is prepared to produce a three-dimensional representation of the trends in electronegativity, atomic diameter, and ionization energy shown by the periodic arrangement of selected atoms.

Procedure

Wear safety goggles.

1. Prepare a board as shown in Figure 1. Cut out a portion of a periodic chart or draw one and paste it to the board.
2. With a sharp instrument, make a hole in the center of each square to represent an atom on the chart.
3. Cut toothpicks in various lengths to represent the magnitude of the property to be illustrated. A convenient scale is

   electronegativity: 1 cm = 1 electronegativity unit
   atomic diameter: 1 cm = 1 angstrom
   ionization energy: 1 cm = 6 eV

   Refer to the periodic chart in Appendix 1 for correct values. Add 0.5 cm for the depth of the hole.
4. Place the toothpicks in the appropriate positions in the chart. Figure 1 shows toothpicks cut to represent the electronegativity of one period and one group of elements.

Teaching Tips

NOTES

1. Colored, round toothpicks work best. Use a different color for each property.
2. You can glue the toothpicks in place for a permanent classroom display.
3. Transitional elements are omitted because properties show little change from element to element.

QUESTIONS FOR STUDENTS

1. What is the general trend in electronegativity across a row of elements? How can you account for this?
2. What is the general trend in electronegativity down a column of elements? How can you account for this?
3. According to the model, which two elements should form the strongest ionic bond?

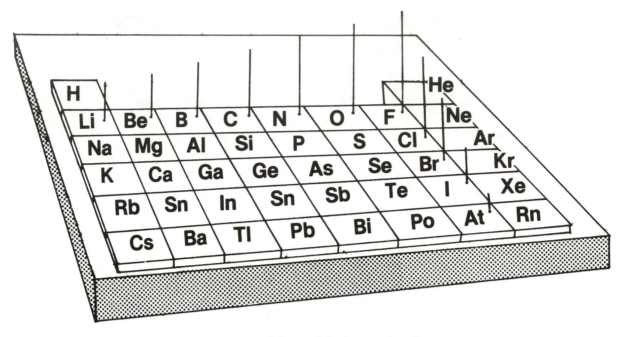

Figure 1. Board for model of properties of atoms.

Gases

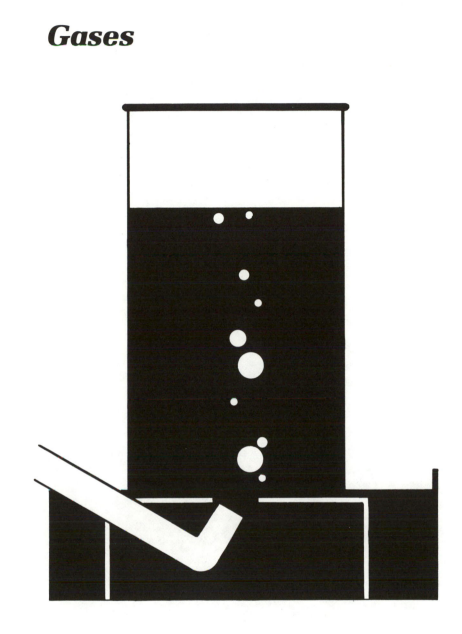

2.   Gas Densities

Large bubbles are formed by bubbling laboratory gas through a soap solution. As the bubbles rise in the air they are ignited with a candle attached to a meter stick. This produces a spectacular effect.

Procedure

Wear safety goggles; have your assistant wear goggles also. Have a fire extinguisher nearby.

1. Either obtain a small bottle of bubble solution at the toy store or prepare a detergent solution. Place the solution in a large beaker.
2. Carefully blow bubbles in the solution by using a hose attached to the laboratory methane gas outlet. A small funnel or thistle tube works well as a pipe.
3. Have an assistant standing by with a birthday candle taped to the end of a meter stick.
4. As the bubbles rise in the air, ignite them with the candle.

Solution

Make the detergent solution by adding 2 heaping spoons (about 70 g) of detergent, 120 mL of glycerin, and a few drops of corn syrup to 120 mL of water.

Teaching Tips

NOTES

1. This demonstration illustrates that methane gas has a density less than that of air.
2. You may need a little practice to get good bubbles. As the bubbles form, gently shake them loose from the funnel and they will float upward.
3. Do not ignite bubbles before they dislodge and float upward.
4. Because it is difficult to produce a bubble larger than 10–15 cm by this method, the fireball will not be large. Do not allow the bubble to reach the ceiling before it is ignited.

QUESTIONS FOR STUDENTS

1. This demonstration works well with methane gas. Would it also work with propane? Butane?
2. Would it work with hydrogen?

3.   Properties of Gases: Pressure and Suction

A deflated balloon is placed between the mouths of two cups. As the balloon is inflated, the cups adhere to the balloon and will not fall off as the balloon is passed around the class.

Procedure

Wear safety goggles.

1. Obtain several balloons that will easily inflate to 6–8 in.
2. Hold two cups with the deflated balloon between them (you may want a student to assist you).
3. Blow into the balloon and inflate it with one breath.
4. Hold up the balloon with the attached cups and ask for an explanation.

Reaction

As the balloon is inflated, air pressure forces the sides of the balloon against the cups with such force that the cups adhere to the balloon. This sequence and the friction between the balloon and the rim of the cup create a suction effect on the cups.

Teaching Tips

NOTES

1. Cheap porcelain cups work well. If you are short on confidence, paper and styrofoam cups work just as well.
2. Try using containers of various sizes and shapes.
3. Pipetting a solution or drinking through a straw are examples of this phenomenon.

QUESTIONS FOR STUDENTS

1. Can you give other examples of this phenomenon?
2. Explain this phenomenon on a molecular basis.

4. Temperature and Pressure Relationships

A large flask containing an inflated balloon is shown to the class.

Procedure

Wear safety goggles.

1. Select an Erlenmeyer flask with a narrow mouth. Place about 5 mL of water in the flask.
2. Heat the flask until the water boils down to a volume of about 1 mL.
3. Remove the flask from the heat, hold it with a towel, and immediately place the open end of a colored balloon over the mouth of the flask. Be careful with the hot flask.
4. Observe the effect as the flask cools. The balloon will be sucked into the flask.
5. To remove the balloon, heat the flask.

Reaction

As the flask cools, the vapor pressure inside the flask decreases. Because the pressure outside the flask is greater, the balloon is sucked (pushed) into the flask.

Teaching Tips

NOTES

1. You will get better results if you partially inflate the balloon before attaching it to the hot flask.
2. Use this example as an inquiry demonstration: Show the flask to your students and ask for possible explanations. See if anyone notices the small amount of water in the flask.
3. Party balloons work best because they have a large opening.

QUESTIONS FOR STUDENTS

1. Why is it necessary to boil the water?
2. Why is the balloon sucked into the flask?
3. Would another liquid work just as well?
4. How can you remove the balloon without breaking the flask?

5.   Solubility of a Gas: The Ammonia Fountain

The setup shown in Figure 2 is displayed. The flask is filled with ammonia gas. The beaker is filled with water. When a small amount of water is squirted into the flask from the dropper, water rises in the glass tube and is sprayed into the flask as a pink fountain. The fountain continues until the flask is almost filled.

Procedure

Wear safety goggles. Surround the apparatus with a safety shield.

1. Set up the apparatus as shown in Figure 2. Notice that the tube in the flask has a fine tip. The flask must be dry.
2. Fill the beaker three-fourths with water and add a few drops of phenolphthalein indicator.

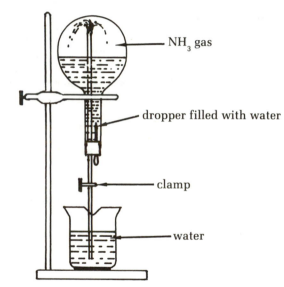

Figure 2. Setup for the ammonia fountain.

3. Fill the dropper with water.
4. Fill the flask with ammonia by the following process:
   a. Place a spatula (7–8 g) of ammonium chloride and a spatula (7–8 g) of sodium hydroxide in a large, dry test tube. Sodium hydroxide is corrosive. Handle it with care. Use gloves.
   b. Gently heat the tube in a hood and direct the ammonia produced into the dry flask. Dispel all air from the flask and collect a full flask of ammonia gas.
5. Place a clamp on the tube inserted in the stopper and stopper the flask.
6. Place the flask upside down in the ring on a ring stand.
7. Place the tube extending from the flask in the beaker. The end of the tube should be just above the bottom of the beaker.
8. When you are ready to begin the demonstration, remove the clamp.
9. Begin the reaction by squirting water from the dropper into the flask.
10. A fountain will result as the water from the beaker is sprayed into the flask.

Reactions

1. Production of ammonia:

$$NH_4^+ \text{ (aq)} + \text{excess } OH^- \text{ (aq)} \rightleftharpoons NH_3 \text{ (g)} + H_2O \text{ (l)}$$

2. The solubility of ammonia gas in water is so great that most of the ammonia immediately dissolves in the water from the dropper. This produces a partial vacuum in the flask. Because of the difference in pressure, water from the beaker rises and enters the flask.
3. The basic ammonium hydroxide produced reacts with the phenolphthalein indicator and turns the solution red.

Teaching Tips

NOTES

1. You can also produce ammonia gas in the flask by adding a small amount of concentrated ammonium hydroxide and gently warming the flask to produce ammonia.
2. If you place about 20 mL 0.1 M HCl in the beaker and add litmus paper rather than phenolphthalein as an indicator, the red solution will become blue as it enters the flask.
3. HCl is also soluble in water. You can make HCl by the following process: Use a face shield and gloves. Gently heat a test tube containing sodium chloride and concentrated sulfuric acid, and direct the HCl produced into a dry flask. If you use litmus paper as an indicator, place a small amount of ammonium hydroxide in the beaker. If you use methyl violet indicator, you will produce yellow, green, blue, and violet solutions. Use only water and indicator in the flask.
4. Experiment with other indicators. Methyl orange works well.
5. If the demonstration does not work, it is probably because the flask was not dry or the flask was not completely filled with ammonia.

QUESTIONS FOR STUDENTS

1. Why will this reaction not work if the flask is wet?
2. Why is ammonia not collected by displacing water, as one would collect oxygen?
3. What other gases show great solubility in water?
4. Explain the reaction used to produce ammonia gas.

6. Preparation of Oxygen Gas from Laundry Bleach

Oxygen gas is prepared by catalytic decomposition of laundry bleach. The gas is collected by displacement of water.

Procedure

Wear safety goggles. Do this demonstration in a hood.

1. Set up a gas-collecting apparatus. A large filtering flask with a hose connected to the side arm and leading into a shallow dish filled with water works well. Collect the gas by displacing water from filled test tubes.
2. Place 100 mL of fresh laundry bleach in the filtering flask.
3. Add approximately 5 g of cobalt(II) nitrate, $Co(NO_3)_2$, to the flask.
4. Stopper the flask quickly and swirl gently to mix contents.
5. Oxygen gas will be produced. Before collecting the gas, displace all the air in the flask and test tubes.
6. Ignite a glowing splint to show the presence of oxygen gas.
7. Heat steel wool and place it in one of the test tubes.
8. Dilute the bleach and flush it down the sink. Wrap the black precipitate in paper towels and place it in the solid waste container.

Reaction

$$2ClO^- \text{ (aq)} \xrightarrow{\text{catalyst}} O_2 \text{ (g)} + 2Cl^- \text{ (aq)}$$
$$\text{hypochlorite}$$

Teaching Tips

NOTES

1. Laundry bleach contains sodium hypochlorite as the active ingredient.
2. The black precipitate is probably an unstable oxide, such as Co_2O_3, that decomposes to form oxygen, then recombines with the hypochlorite ion.
3. Because the rate of this reaction varies with temperature, it can be used to illustrate chemical kinetics:
   a. Add 3 mL of 0.2 M cobalt nitrate to 15 mL of bleach.
   b. Vary the temperature and compare the rates of reaction.

QUESTIONS FOR STUDENTS

1. Record your observations.
2. Write the equation for the reaction.
3. Why did we collect the gas by displacement of water?
4. Why did the splint burst into flame?
5. Why might this reaction rate vary with temperature?

7. Preparation of Chlorine Gas from Laundry Bleach

Chlorine gas and chlorine water are prepared by reacting laundry bleach with hydrochloric acid. The bleaching property of chlorine is shown.

Procedure

Wear safety goggles. Do this demonstration in a hood.

1. Prepare a gas-generating system by connecting a rubber tube to the side arm of a large filtering flask.
2. Place 30 mL of laundry bleach in the flask.
3. Add 5 mL of HCl. Stopper the flask and swirl it quickly.
4. Collect chlorine gas by upward displacement of air in several test tubes. Stopper the tubes.
5. Test for the presence of chlorine by the usual methods (e.g., ability to bleach noncolorfast fabric or dyes).

Reaction

$$ClO^- (aq) + Cl^- (aq) + 2H^+ (aq) \rightarrow Cl_2 (g) + H_2O (l)$$
hypochlorite

Solution

The HCl concentration is 1.0 M (see Appendix 2).

Teaching Tips

NOTES

1. Avoid breathing the chlorine directly.
2. This method is an easy way to prepare chlorine water.
3. Because bleach is prepared commercially by bubbling chlorine gas through sodium hydroxide, this demonstration is essentially the reverse of this reaction:

$$Cl_2 (g) + 2OH^- (aq) \rightarrow ClO^- (aq) + Cl^- (aq) + H_2O (l)$$

4. Place a small piece of cloth soaked in turpentine in one of the tubes. Caution: This will suddenly produce a large amount of soot.
5. Bleach uses nascent oxygen to remove color from fabrics and to kill bacteria.

QUESTIONS FOR STUDENTS

1. Why was this gas not collected by the displacement of water?
2. What property of the gas allows us to collect it by the upward displacement of air?
3. Write the reaction for the preparation of chlorine gas.
4. How does bleach do what it does?
5. What happened with the turpentine? Was this a spontaneous reaction?

8.   Diffusion of Gases

A plug of cotton dipped in HCl is inserted in the open end of a graduated cylinder. The time required for a color change to occur in a strip of pH paper at the other end of the cylinder is noted. The procedure is repeated with a cotton plug dipped in aqueous ammonia. From these observations, Graham's law of diffusion is checked.

Procedure

Wear safety goggles, gloves, and a face shield when you handle concentrated HCl.

1. Clean and dry two 100-mL graduated cylinders.
2. With a glass rod, insert a moist piece of blue litmus paper in one cylinder. Push the paper to the bottom of the cylinder and see that it sticks to the bottom.
3. Place the cylinder on its side, making sure that it is level.
4. Carefully dip a small piece of cotton in concentrated HCl and place it just inside the opening of the cylinder.
5. Immediately place a piece of plastic wrap tightly over the opening of the cylinder.
6. As soon as the cotton plug is inserted, have a student assistant begin timing. Record the time required for the HCl gas to travel the length of the cylinder and cause the blue litmus paper to turn red.
7. Repeat the demonstration with red litmus paper and a cotton plug dipped in concentrated aqueous ammonia in the other graduated cylinder.

Calculations

1. The data collected from this demonstration will be used to check Graham's law of diffusion of gases: The rate of diffusion of a gas is inversely proportional to the square root of its molecular mass.
2. Determine the rate: Divide the distance the gas traveled (length of the cylinder in centimeters) by time (seconds).
3. Determine the ratio of the experimental diffusion rate: Divide the rate of diffusion of hydrogen chloride gas by the rate of diffusion of ammonia gas.
4. Check your answer against the theoretical ratio:

$$\frac{R_{NH_3}}{R_{HCl}} = \sqrt{\frac{M_{HCl}}{M_{NH_3}}} = 1.46$$

Teaching Tips

NOTES

1. Graduated cylinders are used because they are readily available and provide a short distance for the gas to travel.
2. An alternate method is to use a long piece of glass tubing. Simultaneously insert a cotton plug soaked in HCl in one end of the tube and a cotton plug soaked in aqueous ammonia in the other end. Note the appearance of a white ring of ammonium chloride in the tube. Measure the distance from each end of the tube to this ring, and use this number to calculate the rate. (The white ring is often difficult to detect.)

3. Graham's work in 1832, which became Graham's law, involved effusion, rather than diffusion of gases. The ratio is the same for effusion and diffusion, but it arises in two different ways: In effusion, the molecular flux is directly proportional to molecular speed, and thus inversely proportional to the square root of molecular mass. In diffusion, the molecular flux is inversely proportional to molecular momentum, $m\bar{c}$ (or mv), and thus is also inversely proportional to the square root of the molecular mass.

4. This is a good place to discuss experimental error.

QUESTIONS FOR STUDENTS

1. Why did the ratio calculated from the data collected in this demonstration differ from the theoretical ratio?
2. What is the relationship between the mass of gas and its rate of diffusion?
3. Why is it necessary to use dry cylinders for the demonstration?
4. How would the calculated ratio be affected if the cylinders were not level during the demonstration?

9.   Production of a Gas: Acetylene

A small amount of water and calcium carbide is placed inside a rubber balloon. The balloon is tied shut and soon begins to expand as a result of the production of acetylene gas. The expanded balloon is tied to a meter stick and held near a burner flame. Caution: A loud explosion results.

Procedure

Wear safety goggles.

1. With a dropper, squirt 2–3 mL of water into a balloon.
2. Push a piece of calcium carbide (CaC_2) no larger than 5 g through the neck of the balloon. Pinch the balloon to hold it in place while the neck of the balloon is securely tied.
3. Release the balloon and watch as it expands.
4. When the reaction and balloon expansion cease, tape the balloon to the end of a meter stick and hold it near a burner flame. Expect a loud explosion.

Reactions

1. Production of acetylene:

$$CaC_2 \text{ (s)} + 2H_2O \text{ (l)} \longrightarrow Ca(OH)_2 \text{ (s)} + C_2H_2 \text{ (g)}$$

2. Explosion of acetylene:

$$2C_2H_2 \text{ (g)} + 5O_2 \text{ (g)} \longrightarrow 4CO_2 \text{ (g)} + 2H_2O \text{ (g)} + \text{heat}$$

Teaching Tips

NOTES

1. Practice this demonstration to get just the right amounts of water and calcium carbide for maximum effect.
2. We do not recommend that you use a paint can for this demonstration because injury may result from the can lid as it flies off.
3. Twist ties work well to separate water and CaC_2 before mixing.
4. The structure of acetylene contains a triple bond: $HC \equiv CH$.

QUESTIONS FOR STUDENTS

1. What suggestions can you give for storing calcium carbide?
2. Can you draw the structure of acetylene?
3. What are some commercial uses for acetylene?

10.   Molar Volume of Carbon Dioxide

Carbon dioxide is easily and quickly produced by the sublimation of dry ice. By determining the volume of a large self-sealing plastic bag and weighing the amount of carbon dioxide, the molar volume of this gas is easily determined.

Procedure

Wear safety goggles. Use heavy gloves to handle dry ice.

1. Weigh a one-quart self-sealing plastic bag (e.g., Ziploc brand) and a piece of dry ice (about 100–150 g) together to the nearest 0.01 g.
2. Carefully place the dry ice in the bag, press out all of the air in the bag, and zip the bag shut.
3. Observe the bag for 2–3 min until it is inflated with carbon dioxide.
4. Open the bag, press out the gas, and immediately reweigh the bag and the dry ice.
5. Determine the fraction of a mole of dry ice that entered the gaseous phase.
6. Remove the dry ice and fill the bag with water to find the approximate volume of the bag.
7. Using the ratio of volume per mole, find the expected volume per mole (molar volume).

Reactions

The rate of sublimation of dry ice is about 1 g/min.

$$CO_2 \text{ (s)} \rightarrow CO_2 \text{ (g)}$$

Calculations

1. Divide the weight lost by the dry ice by the molecular weight of 1 mol of dry ice to find the fraction of a mole of carbon dioxide produced. Example: If the dry ice lost 2.4 g, then the fraction of a mole is

$$\frac{2.4 \text{ g}}{44 \text{ g/mol}} = 0.054 \text{ mol carbon dioxide}$$

2. Find the molar volume by dividing the volume of the bag by the number of moles. Example: If the amount of water used to fill the bag was 1.3 L, the molar volume is

$$\frac{1.3 \text{ L}}{0.054 \text{ mol}} = 24 \text{ L/mol}$$

Teaching Tips

NOTES

1. The expected volume of 1 mol of any gas at room temperature and pressure is about 24 L. If you wish, you can adjust this to determine the molar volume (22.4 L/mol) at standard temperature and pressure (273 K and 760 Torr).

2. Use gloves or tongs to handle dry ice. Its temperature is –78 °C.

3. Leave the opened bag on the balance as the dry ice sublimes, and you can easily observe the rate of change in mass.

4. Try letting the closed bag fill with gas until it pops. (Warning: The pop is quite loud.) Note that there is little difference between the mass of the filled bag and that of the popped bag.

5. If any moisture condenses on the outside of the bag, wipe it off before reweighing the bag.

6. Because you squeezed essentially all of the air from the bag, you need no correction for the weight of air.

QUESTIONS FOR STUDENTS

1. Why does the bag inflate and pop?

2. What other method might be used to find the volume of gaseous CO_2?

3. What sources of experimental error exist in this demonstration?

4. What is the volume of 1 mole of hydrogen gas? Of neon?

11.   Determining the Molecular Weight of a Gas

A large graduated cylinder is filled with water and inverted in a water-filled trough. A small piece of rubber tubing from a pocket lighter is placed inside the cylinder. When the release button on the lighter is pressed, butane is released, displacing the water in the cylinder. By measuring the volume of water displaced and the weight of the gas, the molecular mass of butane is calculated.

Procedure

Wear safety goggles.

1. Remove the striking mechanism (the flint, wheel, and spring) from a new disposable pocket lighter.
2. Weigh the lighter on a balance. If you plan to use a small cylinder and displace a small volume of water, use an analytical balance. If you use a 500-mL cylinder, you can use a triple-beam balance. Record the weight.
3. Attach a narrow rubber tube to the gas nozzle of the lighter.
4. Fill the largest graduated cylinder you have with water, invert it, and place it inside a trough half-filled with water. Be sure that the cylinder contains no gas bubbles.
5. Place the end of the rubber tube under and inside the cylinder and press the release button on the lighter.
6. Collect enough gas to displace 300–400 mL of water, or the largest volume possible. Move the cylinder so that the water level inside and outside are the same.
7. Read the gas volume directly from the inverted cylinder.
8. Remove the tube from the lighter and reweigh the lighter.
9. From these data, calculate the molecular mass of butane. (Remember to subtract the vapor pressure of water—see Table I—and to record the temperature and pressure.)

Calculations

1. PV equals nRT, where n is grams per molar mass. Thus, PV equals (grams per molar mass) multiplied by RT.
2. Molar mass equals (grams) times RT/PV, where R is the gas constant, 62,400 mL Torr · mol^{-1} · K^{-1}; T is the temperature in Kelvins (273 K = 0 °C); P is the pressure in Torr corrected for the vapor pressure of water; and V is the volume of gas in milliliters. (If you prefer to use L · atm · mol^{-1} · K^{-1}, use $R = 0.0821$ and convert pressure to atmospheres and volume to liters.)

Teaching Tips

NOTES

1. The molecular mass of butane is 58.
2. Now is a good time to discuss experimental error.
3. If the tube on the lighter leaks, your results will be off. Try holding the lighter directly beneath the water-filled cylinder and releasing the gas.

4. Use paper towels to dry the lighter prior to final weighing.
5. In step 6, air pressure outside and inside the cylinder is equal.

QUESTIONS FOR STUDENTS

1. How can you explain the fact that butane is a liquid in the lighter but a gas when it is collected?
2. If you only have a triple-beam balance, why do you need to collect a large volume of gas?
3. Why is it necessary to subtract the vapor pressure of water?
4. How does the vapor pressure of water vary with temperature?

Table I. Vapor Pressure of Water

Temperature (°C)	Pressure (Torr)
15	12.8
16	13.6
17	14.5
18	15.5
19	16.5
20	17.5
21	18.6
22	19.8
23	21.0
24	22.4
25	23.7
26	25.2
27	26.7
28	28.3
29	30.0
30	31.8

12.   The Effect of Pressure on Boiling Point

A syringe is half-filled with water at a temperature below the boiling point. The end of the syringe is closed. When the plunger is pulled, the pressure is decreased, and the water boils.

Procedure

Wear safety goggles.

1. Heat water in a beaker to about 80 °C, well below boiling.
2. Using a large glass or plastic syringe with a short piece of rubber tubing attached, draw about 40–50 mL of hot water into the syringe.
3. Holding the syringe upright, push in the plunger to dispel any air in the syringe.
4. Clamp the tube tightly with a screw-type buret clamp.
5. Holding the syringe with the plunger up, slowly but forcefully pull on the plunger.
6. As the plunger is raised, the pressure on the hot water is reduced and the water boils.

Reaction

The normal boiling point of a liquid is the temperature at which vapor pressure of the liquid equals atmospheric pressure. For water, this temperature is 100 °C. If the atmospheric pressure is reduced, a lower temperature will provide a vapor pressure equal to this pressure and the water will boil at the lower temperature.

Teaching Tips

NOTES

1. Pass the syringe around the class. Caution students not to pull the plunger forcefully.
2. Always hold the syringe down and pull the plunger up.

QUESTIONS FOR STUDENTS

1. What is the relationship among temperature, pressure, and boiling?
2. Why do bubbles appear at or near the boiling point?
3. Would this demonstration work with another liquid?

13. Marshmallow Inflation: Pressure and Volume Relationship

A large marshmallow is placed on the platform of a vacuum pump (or inside a side-arm suction flask). The dome is placed on the platform of the pump (or the side-arm flask is connected to a water aspirator with a rubber hose). As the pressure inside the container is reduced, the marshmallow becomes larger.

Procedure

Wear safety goggles.

1. Place a large marshmallow on the platform of a vacuum pump.
2. Cover the marshmallow with the glass dome and turn on the pump.
3. Notice that the marshmallow gets larger as the pressure inside the dome decreases.

If you have no vacuum pump, connect a side-arm flask with a rubber stopper to fit to an aspirator on a water faucet. Use heavy-walled tubing for the connection.

1. Place a marshmallow in the flask. (Squeeze it through the small opening.)
2. Place the stopper in tightly, connect it to the aspirator with the hose, turn on the water, and observe as the marshmallow becomes larger.

Reaction

The marshmallow has a large amount of air trapped in tiny cells in its carbohydrate structure. As the air outside the marshmallow is reduced, the pressure from the air inside the marshmallow causes the inside air to expand, causing the marshmallow to become larger.

Teaching Tips

NOTES

1. This demonstration usually works well the first time but generally cannot be repeated with the same marshmallow.
2. Try using several smaller marshmallows.
3. This is a good way to demonstrate the relationship between the volume and pressure of a gas.
4. If you have neither a vacuum pump nor a water aspirator, you can try using a large plastic syringe. Place the marshmallow in the flask, stopper the flask, and insert the syringe through the stopper. Remove the air from the flask in one quick pull of the plunger. The results will not be as dramatic, but the effect will be apparent.
5. Warming a marshmallow will also cause it to expand as we observe when we place a marshmallow in hot cocoa. Observe quickly before the marshmallow melts.
6. The air molecules are moving in rapid, random motion. This causes them to expand the marshmallow.
7. Boyle's law is being demonstrated by the expanding volume of gas at reduced pressure: $P_1V_1 = P_2V_2$.

QUESTIONS FOR STUDENTS

1. Why does the air inside the marshmallow fill a larger volume when some of the air outside the marshmallow is removed?
2. What are the air molecules doing that causes them to expand the marshmallow?
3. What gas law is being demonstrated by this expanding volume of gas at reduced pressure?

14.   A Chemical Pop Gun

A solid and a liquid are added to a large test tube. The tube is immediately stoppered and pointed away from the class. A loud pop results and the stopper is shot across the room.

Procedure

Wear safety goggles, a face shield, and thick gloves.

1. Wrap cloth tape tightly around a large test tube.
2. Place about 10–15 mL of vinegar in the tube.
3. Add a spoon (7–8 g) of sodium carbonate wrapped in a small piece of tissue.
4. Immediately cork the tube and hold it at arm's length. Point it in a safe direction, away from the students.

Reaction

The acetic acid (vinegar) and the carbonate react to form carbon dioxide gas. When confined, the gas exerts enough pressure to force the cork from the tube.

$$Na_2CO_3 \ (s) + 2CH_3COOH \ (aq) \rightarrow 2NaCH_3COO \ (aq) + H_2CO_3 \ (aq)$$

$$H_2CO_3 \ (aq) \rightarrow H_2O \ (l) + CO_2 \ (g)$$

Solutions

Use regular-strength vinegar, which is 5.25% acetic acid.

Teaching Tips

NOTES

1. This simple demonstration illustrates various reactions, including the action of an acid on a base, the relationship between gas pressure and volume, and a double decomposition.
2. With a little experimenting, you will find the right combination of vinegar and carbonate for the maximum effect.
3. If you do not get a loud pop, the cork probably is not fitted tightly enough.

QUESTIONS FOR STUDENTS

1. Why must the cork be firmly fitted in the tube?
2. Write the equation for this reaction.
3. What is the relationship between gas pressure and volume? How does this demonstration show this relationship?

15.   The Methanol Cannon

A corked plastic bottle is clamped into position on a ring stand. An electric spark is applied with a Tesla coil to one of two nails placed through the sides of the bottle. A loud explosion occurs, and the cork is propelled across the room.

Procedure

Wear safety goggles. Place a shield around the apparatus.

1. Prepare the "cannon" by inserting two large nails through the sides of a heavy plastic bottle. Shampoo or juice bottles work well. The points of the nails should be separated by about ¼ in. to provide a gap (Figure 3).

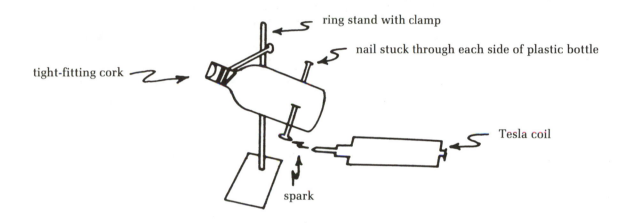

Figure 3. Setup for the methanol "cannon".

2. Add about 1 mL of methanol to the bottle.
3. Shake the bottle to vaporize and distribute the methanol.
4. Place a tight-fitting cork in the mouth of the bottle.
5. Securely fasten the bottle on a ring stand by clamping the neck of the bottle with a clamp attached to a ring stand. Direct the mouth of the bottle up and away from the students.
6. Turn on the Tesla coil and apply a spark from the coil to the head of one of the nails in the bottle.
7. A loud explosion will result, and the cork will be propelled across the room.

Reaction

The spark ignites the methanol vapor. A rapid exothermic reaction, carbon dioxide, and water are produced.

$$4CH_4OH \text{ (g)} + 7O_2 \text{ (g)} \rightarrow 4CO_2 \text{ (g)} + 10H_2O \text{ (g)}$$

Teaching Tips

NOTES

1. You can also use ethanol.
2. Do not use more than 1.0 mL of alcohol. There will probably be enough vapor left for a second reaction, even with this small amount.
3. After a second firing, rinse the bottle with water and dry it before using the cannon again. If you allow the bottle to stand without cleaning for 1–2 days, it will fire again.
4. The blue flame of the reaction is clearly visible in a darkened room.

QUESTIONS FOR STUDENTS

1. Why does such a loud explosion result?
2. Write an equation for the reaction that occurs.
3. Why must the bottle be cleaned before it can be used again?

16.   The Combination of Hydrogen and Oxygen

A balloon is taped to a corner of the demonstration desk. The teacher brings a burning match, attached to the end of a meter stick, to the balloon. The balloon explodes.

Procedure

Wear safety goggles. Place a shield around the balloon.

1. The balloon contains a mixture of hydrogen and oxygen gases.
2. Prepare the hydrogen gas as follows (unless you have a lecture bottle of hydrogen): Place approximately 50 mL of 6 M HCl in a flask (see Appendix 2); add a few pieces of mossy zinc. As soon as gas production is evident, attach a balloon to the mouth of the bottle. Wear gloves and a face shield when you use 6 M HCl.
3. Allow the balloon to inflate as much as possible (about 6 in.).
4. Carefully remove the balloon without losing any of the hydrogen gas and inflate it to approximately double its size by adding oxygen from a tank.
5. Tie the balloon and tape it to the end of the demonstration table away from the students. Clear the area of glassware.
6. Tape a match to the end of a meter stick. Light the match and touch it at arm's length to the balloon.

Reaction

$$2H_2 \text{ (g)} + O_2 \text{ (g)} \longrightarrow 2H_2O \text{ (g)} + \text{heat}$$

Teaching Tips

NOTES

1. You may need to experiment to get just the right explosive mixture.
2. Avoid using concentrated HCl to generate hydrogen gas.
3. You can also perform this demonstration with a large plastic milk jug. Cut the bottom from the jug and fit a one-hole stopper containing the glass tube from a medicine dropper in the mouth of the jug. Fill the jug with hydrogen from a lecture bottle. Hold a match to the top of the glass tube. The hydrogen will burn with a pale blue flame. As the hydrogen is consumed, the flame will get smaller and oxygen will be drawn into the jug through the bottom. When the proper combustion mixture results, the hydrogen will ignite. Place a safety shield around the jug.
4. Hydrogen will burn in pure oxygen to produce temperatures as high as 2888 °C. This oxyhydrogen torch will easily cut thick sheets of metal.
5. The *Hindenburg* airship exploded in 1937 as a result of a spark igniting hydrogen gas.

QUESTIONS FOR STUDENTS

1. Write an equation for this reaction.
2. Hydrogen has a high heat of combustion when ignited with oxygen. Can you think of any practical uses for this?
3. Compare the chemical properties of hydrogen with those of oxygen.
4. Is hydrogen now used in blimps and dirigibles? Why?

Solubility and Solutions

17.   Precipitate Formation: White

Two colorless solutions are mixed. In 1–2 min, a copious, flaky white precipitate forms.

Procedure

Wear safety goggles and disposable plastic gloves. These salts are toxic.

1. Place 180 mL of solution A in a large beaker.
2. Add 40 mL of solution B. Do not stir.
3. After a few minutes a heavy precipitate will form.
4. Place the beaker on the demonstration desk and note the continued precipitation over the next 5–10 min.

Reaction

$$2K(SbO)C_4H_4O_6 \text{ (aq)} + BaCl_2 \text{ (aq)} \rightarrow Ba[(SbO)C_4H_4O_6]_2 + 2KCl \text{ (aq)}$$
$$\text{precipitate}$$

Solutions

1. Solution A, antimony potassium tartrate: Dissolve 27 g of $K(SbO)C_4H_4O_6$ in 180 mL of distilled water.
2. Solution B, barium chloride: Dissolve 10 g of $BaCl_2$ in 40 mL of distilled water.
3. You may need to heat both solutions gently.

Teaching Tips

NOTES

1. Barium chloride is toxic. Wear gloves and avoid contact with this substance.
2. Antimony potassium tartrate is also called potassium antimony tartrate.
3. The precipitate should be collected, washed with alcohol, and dried. Dispose of it according to the directions given in Appendix 5.

QUESTIONS FOR STUDENTS

1. Write the chemical equation for this reaction.
2. What type of reaction does this demonstration show?
3. What causes a precipitate to form when the two solutions are mixed?

18.   Precipitate Formation: Black and White

Two clear solutions are placed in separate beakers labeled 1 and 2. When the contents of beaker 1 are poured into beaker 2, a white precipitate immediately forms. When the process is repeated but this time the contents of beaker 2 are poured into beaker 1, a black precipitate is formed. Two different results are produced by mixing the same solutions.

Procedure

Wear safety goggles and disposable gloves. Mercury chlorine is toxic.

1. Place 10 mL of $SnCl_2$ solution in beaker 1.
2. Place 90 mL of $HgCl_2$ solution in beaker 2.
3. Rapidly pour the contents of beaker 2 into beaker 1.
4. Note the formation of a white precipitate.
5. Place 10 mL of $HgCl_2$ solution in beaker 1.
6. Place 90 mL of $SnCl_2$ solution in beaker 2.
7. Rapidly pour the contents of beaker 2 into beaker 1.
8. Note the formation of a black precipitate.
9. Dispose of the solid waste according to directions given in Appendix 5.

Reactions

The first reaction is

$$2Hg^{2+} (aq) + Sn^{2+} (aq) + 2Cl^- (aq) \rightarrow Hg_2Cl_2 (s) + Sn^{4+} (aq)$$
$$\text{white precipitate}$$

The second reaction is

$$HgCl_2 (s) + Sn^{2+} (aq) \rightarrow Hg (s) + Sn^{4+} (aq) + 2Cl^- (aq)$$
$$\text{black}$$
$$\text{precipitate}$$

Solutions

1. The $SnCl_2$ solution is 0.1 M: 19 g of $SnCl_2$ per liter of water.
2. The $HgCl_2$ solution is 0.1 M: 27.2 g of $HgCl_2$ per liter of water.
3. Add a drop or two of concentrated HCl to each solution to prevent the formation of metal–hydroxy complexes.

Teaching Tips

NOTES

1. The $SnCl_2$ solution must be made fresh prior to the demonstration.
2. Always pour the larger volume into the smaller volume.
3. This demonstration was developed some time ago by our colleague, Jay A. Young. We are pleased to acknowledge his contribution.

QUESTIONS FOR STUDENTS

1. Write the chemical equations for the two reactions.
2. What are the formula and name of the white precipitate?
3. What are the formula and name of the black precipitate?
4. Why do they form differently?
5. Do the different volumes of solutions have an effect on the reaction?

19.   Precipitate Formation: Blue

A small amount of wine-colored solution is added to a clear solution. A blue precipitate forms.

Procedure

Wear safety goggles.

1. Place approximately 200 mL of clear limewater in a beaker.
2. Add a few drops of cobalt nitrate solution from a dropper.
3. Note the immediate formation of a blue precipitate.

Reaction

$$Co(NO_3)_2 \text{ (aq)} + Ca(OH)_2 \text{ (aq)} \rightarrow Ca^{2+} \text{ (aq)} + 2NO_3^- \text{ (aq)} + Co(OH)_2 \text{ (s)}$$
blue precipitate

Solutions

1. The limewater, $Ca(OH)_2$, is a saturated solution. Let it stand overnight and pour off the clear supernate.
2. The concentration of cobalt nitrate, $Co(NO_3)_2$, is not critical; try various amounts.

Teaching Tips

NOTES

1. This method is a good demonstration to show students that not all precipitates are white.
2. This demonstration projects well; use Petri dishes on an overhead projector.

QUESTIONS FOR STUDENTS

1. Write the equation for the reaction.
2. What do you know about the general solubility of hydroxides?
3. What is the blue precipitate?
4. Could you do this demonstration by adding cobalt ion to another hydroxide such as 0.5 M NaOH?

20. Effect of Temperature on Solubility

A large stoppered test tube containing a clear solution is placed in a mixture of salt and ice, and pink crystals separate out. The tube is then heated and the solution becomes clear again. Upon further heating, a white precipitate forms.

Procedure

Wear safety goggles.

1. Prepare a mixture of salt and ice to provide a low-temperature solution.
2. Place the test tube in the mixture. Note the formation of pink crystals.
3. Heat the tube to 27 °C. Note that the solution becomes clear.
4. Continue to heat the tube above 27 °C. Note the formation of a white precipitate.

Reactions

1. The pink crystals are $MnSO_4 \cdot 7H_2O$.
2. The white precipitate is $MnSO_4 \cdot H_2O$.

$$MnSO_4 \cdot 7H_2O \text{ (s)} + \text{heat} \rightleftarrows Mn^{2+} \text{ (aq)} + SO_4^{2-} \text{ (aq)} + \text{heat} \rightleftarrows MnSO_4 \cdot H_2O \text{ (s)}$$

pink   clear   white
precipitate   solution   precipitate

Solutions

Prepare the solution in the test tube as follows:

1. Add a few drops of concentrated sulfuric acid to 100 mL of water at 27 °C.
2. Add either 70 g of $MnSO_4$ or 130 g of $MnSO_4 \cdot 7H_2O$.
3. Decant the clear solution into a test tube and seal the tube.

Teaching Tips

NOTES

1. This demonstration shows that solubility generally increases with temperature but that this rule does have exceptions.
2. Have your students determine the temperature of the salt and ice mixture.
3. Any time mercury thermometers are used, be prepared to clean up spilled mercury (Appendix 5).

QUESTIONS FOR STUDENTS

1. This reaction is temperature-dependent. Why?
2. Do the precipitates contain water of hydration?
3. What happens at a higher temperature?
4. Write the formulas for the two precipitates.
5. Can you think of other compounds that might behave the same way?

21.   Negative Coefficient of Solubility: Calcium Acetate

A clear solution is heated and a precipitate forms. When the solution is cooled, the precipitate dissolves.

Procedure

Wear safety goggles.

1. Place 150 mL of calcium acetate solution in an Erlenmeyer flask and heat the flask. Record the temperature.
2. Note that a precipitate begins to form at approximately 80 °C.
3. Remove the flask from the heat and cool it by placing it in a stream of cold water. Record the temperature.
4. Notice that the solid goes back into solution as the temperature decreases.

Solution

The calcium acetate solution is saturated (about 40 g/100 mL of water).

Teaching Tips

NOTES

1. This demonstration shows students that most rules in chemistry have exceptions.
2. The difference in solubility at the two temperature extremes is about 15 g/100 g of water: At 0 °C, the solubility is 44–52 g/100 g of water; and at 100 °C, the solubility is 36–45 g/100 g of water.

QUESTIONS FOR STUDENTS

1. What appears to happen?
2. Is this what you would normally expect? What usually happens?
3. Does water of hydration enter into the results?
4. Devise a model for the expected results.

22.   Supersaturation and Crystallization

A clear solution is dripped over a few crystals on a laboratory bench. As the solution is dripped, a crystal matrix is formed, and a tall crystal column is produced.

Procedure

Wear safety goggles.

1. Clean off an area of the laboratory bench.
2. Place a few small crystals of sodium acetate on the clean area.
3. Slowly drip the solution over the crystals.
4. Crystallization begins and a tall column of crystals forms.

Reactions

A supersaturated solution contains more than the normal saturation quantity of a solute. When a small crystal of solid solute is added, crystallization of the excess solute results, and an equilibrium appropriate to the lower temperature is restored.

$$CH_3COONa \text{ (s)} \rightleftharpoons Na^+ \text{ (aq)} + CH_3COO^- \text{ (aq)}$$

Solutions

Prepare the sodium acetate solution as follows:

1. Place 50 g of sodium acetate trihydrate in a small flask.
2. Add 5 mL of water and slowly warm the flask.
3. Swirl the flask until the solid dissolves completely. If any solid remains on the neck or sides of the flask, wash it down with a small amount of water.
4. Remove the flask from the heat, wrap it in aluminum foil, and allow it to cool at room temperature.

Alternate Method

1. Clean and dry a small Erlenmeyer flask and a stopper that fits it.
2. Fill the flask with solid sodium acetate trihydrate.
3. Slowly heat the flask on a hot plate until it completely liquefies. Heat the solid until it just melts; do not boil it.
4. Using a wash bottle, carefully rinse the neck of the flask with a small amount of water.
5. Insert the stopper and allow the flask to cool to room temperature.
6. Remove the stopper when the flask has cooled and carefully add only one crystal of solid sodium acetate trihydrate to the flask.
7. Observe. You can reheat the flask and use the same material over and over.

Teaching Tips

NOTES

1. This demonstration is a nice variation of the usual reaction in which a solution is seeded by adding a small crystal to a supersaturated solution.
2. Practice to get just the right touch necessary to form a tall column. The solid sodium acetate can be reused.
3. Sodium acetate is in the form of a trihydrate.
4. If you fill a clean buret with the saturated solution, you can form a tall column of crystal by dripping the solution from the buret tip onto a few crystals on the laboratory bench.

QUESTIONS FOR STUDENTS

1. What has happened?
2. Why did the column build upward?
3. What reaction was taking place? Is this a chemical or a physical reaction?
4. Why was the flask wrapped in foil and cooled slowly?

23.   The Silicate Garden

A few small, colored crystals are added to a solution in a large jar or bowl. In a few seconds, large, plantlike growths extend from the crystals.

Procedure

Wear safety goggles and disposable gloves.

1. Select a wide-mouth container (a small fish bowl works nicely) and fill it to within 1 in. of the top with sodium silicate solution.
2. Drop in 3–4 crystals the size of a match head.
3. Notice the growth of the crystals within a few seconds.

Reactions

1. When metal salts are added to a silicate solution, insoluble silicates are formed.
2. When the salts are placed in the sodium silicate solution, a semipermeable membrane is formed around the salt. Because the concentration is greater inside the membrane, water enters the membrane to dilute the concentrated solution. This effect is called *osmosis.*

   Osmosis causes the membrane to break. It breaks upward because the pressure of water on the sides of the crystal is greater than that on the top. This process is repeated as a new membrane forms, and an upward growth of the crystal garden results.

Solutions

1. The sodium silicate (also called water glass) is a diluted solution with a specific gravity of about 1.10. A dilution of 1 part of sodium silicate to 4 parts of water is a good approximation.
2. Use the following crystals to get a variety of colors in the garden: iron(III) chloride, brown; cupric chloride, bright green; cobaltous nitrate, dark blue; manganous nitrate, white; and zinc sulfate, white.

Teaching Tips

NOTES

1. Covering the bottom of the container with a thin layer of sand will prevent the crystals from sticking to the bottom of the glass.
2. You can keep this as a permanent display. If the solution becomes cloudy after a few days, carefully remove it and replace it with water.
3. This method is an excellent way to demonstrate osmosis.

QUESTIONS FOR STUDENTS

1. Why do the crystal trees grow upward?
2. A large crystal appears to grow from a small crystal. Explain this.
3. How is osmosis involved in this phenomenon?
4. Is the growth continuous? If not, explain.
5. Would crystals of other chemicals (e.g., calcium chloride, copper nitrate, potassium ferrocyanide) produce such growths?

24.   The Effect of Temperature on a Hydrate: Pink to Blue

A beaker containing a pink solution is heated on a hot plate. As the solution becomes warm, the color changes to blue. The beaker is removed from the heat and allowed to cool. The pink color returns.

Procedure

Wear safety goggles and disposable gloves. Cobalt chloride is toxic. Ethyl alcohol is flammable; extinguish all flames.

1. Place approximately 10 mL of 95% ethyl alcohol in a glass Petri dish. Let it sit until it reaches room temperature.
2. Add a pinch of cobalt(II) chloride and stir until it is dissolved. The solution should be pink. Place the dish on an overhead projector and project the pink color on a screen.
3. Place the dish on a hot plate and warm the solution. The solution will turn blue.
4. Remove the dish from the hot plate and replace it on the projector. As it cools, the solution becomes pink again.

Reactions

1. Cobalt(II) chloride is in the form of a hydrate. When the solution is heated, the blue anhydrous form is produced.

$$\underset{\text{pink}}{Co(H_2O)_6^{2+}} + 4Cl^- \rightleftharpoons \underset{\text{blue}}{CoCl_4^{2-}} + 6H_2O \ (l)$$

2. When the solution is cooled, water in the solution recombines with the cobalt(II) chloride. The pink hydrate is again formed.

Solution

Use 95% ethyl alcohol, not pure ethyl alcohol. You may need to add water to 95% alcohol until you get a pink color in step 2.

Teaching Tips

NOTES

1. This reaction is essentially the same as that produced in weather indicators: Strips of blue anhydrous cobalt(II) chloride paper turn pink when the humidity is high.
2. Demonstrations 25, 47, and 48 offer variations of this reaction.

QUESTIONS FOR STUDENTS

1. Why is 95% alcohol used in this demonstration?
2. Show the reaction between the pink and blue forms of this compound.
3. What part does heat play in this reaction?

25. Cobalt Complexes: Changing Coordination Numbers

A small amount of cobalt chloride solution is added to test tubes containing varying amounts of ethyl alcohol. The solutions in the tubes turn red, violet, and blue; these color changes represent changes in the coordination number of cobalt chloride from 6 to 4.

Procedure

Wear safety goggles and disposable gloves. Cobalt chloride is toxic. Ethyl alcohol is flammable; extinguish all flames.

1. Fill a small test tube one-half with ethyl alcohol. Fill a second tube one-fourth, and a third one-eighth with ethyl alcohol.
2. Add a dropper of cobalt chloride solution to each tube.
3. Note the change in color in each tube.

Reactions

1. The general reaction is:

$$Co(H_2O)_6^{2+} + 4Cl^- \rightleftharpoons CoCl_4^{2-} + 6H_2O \text{ (l)}$$

pink   blue

2. As molecules of ethyl alcohol replace the water molecules in the coordination positions of the complex ion, the color of the solution changes. When one alcohol is attached, the color is red; when two are attached, the color is violet; and when three are attached, the solution is a deep blue.
3. The coordination number of the cobalt(II) ion changes from 6 to 4.

Solutions

1. The ethyl alcohol is 100%.
2. The cobalt(II) chloride solution is 2M: Dissolve 2.6 g of $CoCl_2$ in 10 mL of water.

Teaching Tips

NOTES

1. You can also demonstrate the color change from the hydrated to the dehydrated form with a "magic" writing reaction:
   a. Write a message on white paper with the pink hydrated solution.
   b. When the paper is dry, gently warm it. Dehydration of the salt will produce the blue dehydrated form, and the message will appear.
2. The blue $CoCl_4^{2-}$ ion is tetrahedral (see structure).

QUESTIONS FOR STUDENTS

1. What are coordination numbers?
2. How do the coordination numbers change with the addition of alcohol molecules?

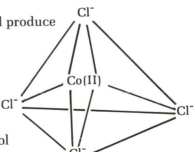

3. How are complex ions with various coordination numbers characterized by colors?

4. Show the reaction for the formation of the blue anhydrous complex.

26. Polar Properties and Solubility

A large graduated cylinder is filled with, seemingly, a single liquid. A deflagrating spoon containing a few crystals of iodine is lowered into the top part of the liquid. The solution becomes dark. The spoon is lowered midway into the cylinder; no color results. The spoon is then lowered to the bottom of the cylinder and violet is produced; this activity produces three distinct layers of liquids in the cylinder.

Procedure

Wear safety goggles and use a hood. Extinguish all flames.

1. Pour 30 mL of trichlorotrifluoroethane (TTE) into a 100-mL graduated cylinder.
2. Slowly add 30 mL of water; allow it to run down the side of the cylinder to avoid mixing.
3. Add 30 mL of petroleum ether; again, allow it to run down the side of the cylinder. Prepare this solution in advance so that the cylinder appears to contain only one liquid. Cover the top of the cylinder.
4. Place a few crystals of iodine in a deflagrating spoon. Slowly lower the spoon into each liquid layer.

Reactions

1. Iodine, TTE, and ether are nonpolar substances and thus dissolve in each other. Therefore, the characteristic colors of iodine in solution are produced.
2. Water is a polar substance and does not readily dissolve a nonpolar substance, such as iodine.

Solution

TTE is trichlorotrifluoroethane. You can also use dichloromethane.

Teaching Tips

NOTES

1. Dispose of the TTE according to directions in Appendix 5.
2. If you don't have a deflagrating spoon, drop a few large pieces of iodine into the cylinder. They will partially dissolve as they sink to the bottom, producing the same effect.
3. This demonstration is an excellent way to introduce molecular polarity.
4. A tincture is a solution of iodine in alcohol

QUESTIONS FOR STUDENTS

1. What prevents the three liquids from mixing?
2. Why are the colors in the top and bottom layer different?
3. What is a tincture?
4. What will happen if you cover the top of the cylinder and allow all three layers to mix?

27.   Hydration of Cu(II) Ion

A small sample of copper(II) bromide dissolved in alcohol produces a dark brown tincture. However, when water is added, a blue–green solution results.

Procedure

Wear safety goggles. Be careful: Ethyl alcohol is flammable. Extinguish all flames.

1. Dissolve a few crystals of copper(II) bromide ($CuBr_2$) in 25 mL of ethyl alcohol. Notice that the resulting tincture is dark brown.
2. Slowly add water to the alcohol solution. Notice the immediate appearance of the blue–green of the hydrated copper(II) ion.

Reactions

Like many other metal ions, Cu^{2+} from $CuBr_2$ reacts with water to produce a hydrated copper ion:

$$CuBr_2 \text{ (s)} + 4H_2O \rightarrow [Cu(H_2O)_4]^{2+} + 2Br^- \text{ (aq)}$$
dark brown   blue–green

Solution

Ethyl alcohol is 95%.

Teaching Tips

1. You can also do this by adding crystals of copper(II) bromide to 5 mL of water. Notice the brown solution. When more water is added, the solution becomes blue. If the blue solution is gently heated to remove the water, the concentrated solution becomes brown again.
2. A substance dissolved in alcohol produces a tincture.
3. Hydrated copper(II) ion, like most other metal ions, forms an acidic solution. The first stage in the ionization of the hydrated copper ion may be shown as

$$[Cu(H_2O)_4]^{2+} \rightleftharpoons [Cu(OH)(H_2O)_3]^+ + H^+$$

4. Because the hydration of a metal ion involves the reaction of a salt with water, these are sometimes called *hydrolysis* reactions.
5. The bonding of molecules or ions of a dissolved substance to solvent molecules is generally called *solvation*. The bonding of an ion specifically to water is called *hydration*.

28. Surface Tension of Water: The Magic Touch

Powdered sulfur is sprinkled on the surface of water in a large beaker. The sulfur floats on the surface; however, when it is touched with the finger the sulfur suddenly cascades to the bottom.

Procedure

Wear safety goggles.

1. Sprinkle enough sulfur on the surface of the water in a large beaker to cover the surface lightly. Do not use pieces large enough to sink.
2. Invite several students to touch the surface of the water. Nothing will happen.
3. Dip your finger in dishwashing detergent or some other wetting agent.
4. Now, when the surface of the water is touched, the sulfur particles will suddenly fall to the bottom.

Reaction

1. The high surface tension of water acts somewhat like an elastic membrane stretched across the water and prevents the sulfur particles from sinking.
2. A wetting agent lowers the surface tension of water and allows the particles of sulfur to drop through the surface.

Teaching Tips

NOTES

1. Experiment to find the type and amount of wetting agent that works best for you.
2. A single drop of diluted dishwashing detergent, or a grain of powdered detergent will produce the cascade effect.
3. Other substances, such as pepper, work well also.

QUESTIONS FOR STUDENTS

1. Why are wetting agents used in dishwashers?
2. Wetting agents are used in washing powders to produce suds. How does this work?
3. How does the surface tension of water compare with that of other liquids?
4. Would this demonstration work in alcohol?

29.   The "Lemonade" Reaction

Water is poured from a large beaker into five smaller beakers. Solutions that are light yellow, orange, red, black, and finally yellow are produced. When the solution in the final yellow beaker is poured back into the large beaker, the entire contents of the larger beaker turns yellow; hence, "lemonade".

Procedure

Wear safety goggles.

1. Fill a large beaker or jar with water.
2. Pour water into beaker 1. Note the formation of a pale yellow solution. Pour this solution back into the large beaker, and express disappointment that "lemonade" was not produced.
3. Pour from the large beaker into beaker 2. Note the orange color. Pour this solution back into the large beaker and try again.
4. Pour water into beaker 3 and note the red solution. Pour this solution back into the large beaker.
5. Pour water into beaker 4. A black color is produced. Pour this solution back into the large beaker.
6. Pour water into beaker 5. You have produced "lemonade".
7. When you pour the contents of beaker 5 back into the dark solution in the large beaker, the contents of the large beaker will turn yellow.

Reactions

Dilute $FeCl_3 \cdot 6H_2O$ is light yellow.

$$Fe^{3+} (aq) + SCN^- (aq) \rightarrow Fe(SCN)^{2+}$$
$$\text{red}$$

$$Fe^{3+} (aq) + \text{tannic acid} \rightarrow Fe(III) \text{ tannate (aq)}$$
$$\text{black}$$

$$2Fe^{3+} (aq) + 3C_2O_4^{2-} (aq) \rightarrow Fe_2(C_2O_4)_3 (aq)$$
$$\text{yellow}$$

Excess oxalate reacts with all of the Fe^{3+} in the beaker to form yellow iron(III) oxalate.

Solutions

1. Dissolve 30 g of ferric chloride ($FeCl_3 \cdot 6H_2O$) in 100 mL of water. Place 15 drops of this solution in beaker 1.
2. Dissolve 22 g of ammonium thiocyanate (NH_4SCN) in 100 mL of water. Place 2 drops of this solution in beaker 2 and 10 drops in beaker 3.
3. Prepare a saturated solution of tannic acid. Place 12 drops in beaker 4.
4. Prepare a saturated solution of oxalic acid. Place 10 mL (*not* 10 drops) in beaker 5.

Teaching Tips

NOTES

1. Place the five solutions in dropper bottles; they can be used over a long period of time.
2. Vary the amounts until you find the combination that gives you the best colors.
3. Tannic acid has the formula $C_{76}H_{52}O_{46}$.

QUESTIONS FOR STUDENTS

1. Write the equations for the reaction that occurs when each small beaker is filled.
2. Are all of these reactions redox reactions?
3. Of those that are redox reactions, what was oxidized and what was reduced?
4. Could the order of pourings be changed and the results still be the same? Why or why not?

30. "Magic" Writing Reactions

Three demonstrations are described that involve the sudden appearance of written messages. These reactions are fun and they illustrate various chemical reactions resulting in color changes.

Procedure

Wear safety goggles.

1. Prepare a writing surface by rubbing a piece of cardboard or poster board with dry ferric chloride. Paint it with colorless solutions of the following chemicals to get the color indicated: potassium thiocyanate (KSCN), red; potassium ferro-cyanide [$K_4Fe(CN)_6$], blue; and tannic acid ($C_{76}H_{52}O_{46}$), black.
2. Prepare a writing surface by rubbing poster paper with a mixture of dry potassium ferrocyanide and ferric ammonium sulfate. Paint with a brush dipped in water to produce a blue color.
3. Soak a piece of paper in concentrated potassium thiocyanate solution. When the paper is dry, dip a finger into a dilute solution of ferric chloride and draw a "bloody" picture on the paper.

Reactions

$$SCN^- \text{ (aq)} + Fe^{3+} \text{ (aq)} \rightarrow \underset{red}{Fe(SCN)^{2+}}$$

$$Fe^{3+} \text{ (aq)} + K_4Fe(CN)_6 \text{ (aq)} \rightarrow 4K^+ \text{ (aq)} + Fe^{2+} \text{ (aq)} + \underset{blue}{[Fe(CN)_6]^{3-}} \text{ (aq)}$$

$$Fe^{3+} \text{ (aq)} + \text{tannic acid} \rightarrow \underset{black}{Fe(III) \text{ tannate (aq)}}$$

Teaching Tips

NOTES

1. Make up your own variations.
2. A favorite is the "bloody" finger. Prepare several sheets of paper as in step 1 and store them in a jar. Pretend to prick the end of a finger. Quickly dip the finger in the ferric chloride solution, and draw on the paper with "blood". Wash your hands after this demonstration (and all other demonstrations)!

QUESTIONS FOR STUDENTS

1. Write the chemical equations for the reactions.
2. Are these redox reactions?
3. Can you suggest another "magic" ink?
4. What is the chemistry of your suggestion?

31. "Magic" Writing Reactions: Questions and Answers

A question is written on a large piece of paper using a blue ink. When the paper is warmed with a burner flame, the blue question disappears and the answer appears.

Procedure

Wear safety goggles.

1. Select an appropriate compound and its formula for the demonstration, for example, **sulfuric acid** and H_2SO_4.
2. Using a small paintbrush, write the formula, H_2SO_4, on the bottom half of a large sheet of paper with solution A. Let the paper dry.
3. When you are ready to do the demonstration, write the name of the compound, **sulfuric acid**, on the top part of the paper, using another small brush and the blue solution B.
4. Hold a Bunsen burner or a hair dryer and carefully fan it over the paper.
5. The name of the compound will disappear, and the formula will appear.

Reactions

Disappearing question: The ink used to write the question consists of a solution of copper sulfate and aqueous ammonia in water. This forms the deep blue complex, $[Cu(NH_3)_4]^{2+}$:

$$Cu^{2+} (aq) + 4NH_3 (aq) \rightarrow [Cu(NH_3)_4]^{2+}$$

When this complex is heated, the complex breaks down and the hydrated copper sulfate is dehydrated to form the colorless anhydrous copper sulfate.

Appearing answer: Heat produces a dehydration reaction between the paper and sodium bisulfite. This produces a charred paper and makes the answer visible as dark writing.

Solutions

Solution A: Dissolve 15 g of sodium bisulfite ($NaHSO_3$) in 100 mL of water.

Solution B: Dissolve 2.5 g of copper sulfate pentahydrate ($CuSO_4 \cdot 5H_2O$) in 100 mL of water. Add 30 mL of aqueous ammonia (NH_4OH). Stir the solution.

Teaching Tips

NOTES

1. When ammonia is added to the hydrated copper ion, a series of reactions occurs with ammonia replacing each of the four waters on the copper ion.
2. Try varying the concentration of solutions until you achieve the effect you desire.

3. A solution of cobalt chloride can also be used as invisible ink. When it is warmed, the hydrated complex, $Co(H_2O)_6{}^{2+}2Cl^-$, decomposes, forming the blue cobalt chloride complex. Handle this compound only with gloves. Cobalt chloride is toxic.

4. You can replace solution B with a dilute aqueous ammonia solution that contains a few drops of phenolphthalein indicator. When heated, the red basic solution will become colorless as the ammonia is removed.

5. In a weather indicator, a blue cobalt spot on paper becomes pink when moist. The following reaction occurs:

$$[Co(H_2O)_6]Cl_2 \rightleftharpoons [CoCl_2(H_2O)_2] + 4H_2O$$
$$\text{blue} \qquad\qquad\qquad \text{pink}$$

QUESTIONS FOR STUDENTS

1. What is the chemical reaction that causes the written compound to disappear?
2. What is the chemistry involved in weather indicators?
3. Can you devise a method to use cobalt chloride solution as an invisible ink?

32.   Patriotic Colors: Red, White, and Blue

A liquid is poured from a flask into three beakers. Red, white, and blue solutions are produced.

Procedure

Wear safety goggles and disposable gloves. Lead compounds are toxic.

Prepare the flask and beakers:

1. Flask: Fill with 1.0 M ammonium hydroxide (aqueous ammonia) solution (see Appendix 2).
2. Beakers:

   beaker 1: 5 drops of alcohol and 5 drops of phenolphthalein
   beaker 2: 5–10 drops of saturated lead nitrate solution [30 g of $Pb(NO_3)_2$ in 100 mL of water]
   beaker 3: 5–10 drops of saturated copper sulfate solution (15 g of $CuSO_4$ in 100 mL of water)

3. Fill the three beakers with the solution in the flask to produce red, white, and blue solutions.
4. Dispose of the lead nitrate solution according to the directions given in Appendix 5.

Reactions

1. Ammonium hydroxide (aqueous ammonia) reacts with the indicator to give a red color.
2. A double displacement (metathesis) reaction occurs and lead hydroxide precipitates:

$$Pb(NO_3)_2 \text{ (aq)} + 2NH_4OH \text{ (aq)} \longrightarrow \underset{\text{white}}{Pb(OH)_2 \text{ (s)}} + 2NH_4NO_3 \text{ (aq)}$$

3. A blue complex ion is formed with copper and ammonium ion:

$$Cu^{2+} \text{ (aq)} + 4NH_4OH \text{ (aq)} \longrightarrow \underset{\text{deep blue}}{[Cu(NH_3)_4](OH)_2 \text{ (aq)}}$$

Teaching Tips

NOTES

1. The blue complex ion is tetraamminecopper(II) hydroxide.
2. Adjust the amounts of chemicals to get the desired intensity of color.
3. These colors show up better against a white background.

QUESTIONS FOR STUDENTS

1. Write chemical equations for each of these reactions.
2. Which reaction represents a double displacement?
3. These reactions were produced by adding ammonia. Could they be reversed by adding an acid?
4. Can you design a demonstration that will produce your school colors?

Acids and Bases

33.   Acid–Base Indicators

A few drops of universal indicator are added to a solution. After a few seconds the solution changes color from blue–green to green to yellow to orange to red.

Procedure

Wear safety goggles and disposable gloves. Do this demonstration in a hood.

1. Prepare a solution by adding 250 mL of water to 250 mL of isopropyl alcohol.
2. To this solution add 1 mL of *tert*-butyl chloride.
3. Add approximately 4 mL of universal indicator solution.
4. Add 2–3 drops of NaOH solution.
5. Stir vigorously for 5–10 s.
6. Observe color changes.

Reactions

1. Hydrolysis of *tert*-butyl chloride reduces [OH⁻] and increases [H⁺]:

$$CH_3-\underset{\underset{CH_3}{|}}{\overset{\overset{CH_3}{|}}{C}}-Cl + OH^- \text{ (aq)} \rightarrow CH_3-\underset{\underset{CH_3}{|}}{\overset{\overset{CH_3}{|}}{C}}-OH + Cl^- \text{ (aq)}$$

2. Decreasing the hydroxide concentration causes the indicator to produce the color changes.

Solutions

1. Isopropyl alcohol from the drugstore gives poor results because it is a dilute solution. Try various amounts for best results.
2. The NaOH solution is 1.0 M: Dissolve 0.4 g of NaOH in 10 mL of water.

Teaching Tips

NOTES

1. The reaction takes about 2 min.
2. You can repeat the color cycle by adding a little NaOH to neutralize the HCl produced by the reaction.
3. You can also use *tert*-butyl bromide, but use less (about 0.5 mL). It is more rapidly hydrolyzed than *tert*-butyl chloride, so you can expect color changes to appear sooner. Use the same care with *tert*-butyl bromide that you would with *tert*-butyl chloride.

QUESTIONS FOR STUDENTS

1. What were the initial and final pH levels?
2. What was the trend in pH change?
3. Suggest a mechanism for the reaction.
4. Can the reaction be reversed? How?
5. Can the reaction be repeated? How?

34.   Acid–Base Indicators: Universal Indicator

A violet solution is poured into a large beaker containing dry ice. The solution is poured back and forth between the two beakers. Violet, blue, green, yellow, and finally orange are produced.

Procedure

Wear safety goggles.

1. Place approximately 100 mL of water in a 500-mL beaker.
2. Add a dropper of indicator solution.
3. Add dilute NaOH solution until a dark violet results.
4. Pour this solution into a second large beaker that contains about a cup of crushed dry ice.
5. Pour the solution back and forth between the beakers to produce the various colors.

Reactions

1. Water reacts with CO_2 to produce an acidic solution. As more H_2CO_3 is formed, the acidity increases and the universal indicator changes to different colors.

$$CO_2 \text{ (g)} + H_2O \text{ (l)} \longrightarrow H_2CO_3 \text{ (aq)} \rightleftharpoons H^+ \text{ (aq)} + HCO_3^- \text{ (aq)}$$

2. The following colors are produced at various values:

Color	pH Level
violet	9
blue	8
green	7
yellow	6
shrimp	5
orange	4

A few drops of HCl may be necessary to reach pH 4 and produce orange.

Teaching Tips

NOTES

1. You may need a little practice to get just the right combination of ingredients.
2. You can repeat this demonstration by adding more dilute NaOH solution.

QUESTIONS FOR STUDENTS

1. Why is a range of colors produced in this reaction?
2. Why are sharp, distinct colors produced rather than one color gradually fading into the next?
3. Why was NaOH added at the beginning of the demonstration?

35.   Acid–Base Indicators and pH

Eight graduated cylinders are arranged on the demonstration table in four pairs; each cylinder contains a colored solution. A small lump of dry ice is dropped into each of the cylinders. As reactions proceed in each cylinder, various color changes result. See Table II.

Procedure

Wear safety goggles.

1. Arrange eight graduated cylinders or tall beakers in pairs.
2. Fill each cylinder about three-fourths with water.
3. Add several drops of the following indicators to each pair of cylinders: pair 1, thymolphthalein; pair 2, phenolphthalein; pair 3, phenol red; and pair 4, brom-thymol blue.
4. Add 5–10 mL of aqueous ammonia to each cylinder.
5. If the cylinder containing thymolphthalein does not turn deep blue, add aqueous ammonia until it does. Add the same amount of aqueous ammonia to the other cylinders as well.
6. Add a small piece of dry ice to one cylinder of each pair.
7. Note the changes in color as the CO_2 dissolves and the pH drops. Compare the color in each cylinder to that of the original.

Table II. Color Changes and pH Range for Indicators

Indicator	Color Range	pH Range
thymolphthalein	blue to colorless	10.6–9.4
phenolphthalein	pink to colorless	10.0–8.2
phenol red	red to yellow	8.0–6.6
bromthymol blue	blue to yellow	7.6–6.0

Reaction

As more CO_2 dissolves, the acidity of the solution increases:

$$CO_2 \text{ (g)} + H_2O \text{ (l)} \rightarrow H_2CO_3 \text{ (aq)} \rightleftharpoons H^+ \text{ (aq)} + HCO_3^- \text{ (aq)}$$

Solution

The aqueous ammonia solution is 1 M (see Appendix 2).

Teaching Tips

NOTES

1. The indicators work well within this pH range. You may wish to add or substitute other indicators.
2. Color changes will occur in sequence if the cylinders are arranged as suggested.

QUESTIONS FOR STUDENTS

1. What pH is indicated by the color change in each cylinder?
2. What does the dry ice do?
3. Write a chemical equation to show the reaction.
4. What causes the bubbling (effervescing)?
5. Does the bubbling action have any effect on the reaction?

36. Acid–Base Indicators: A Voice-Activated Chemical Reaction

A flask containing a colored solution is prepared. Each student is invited to remove the stopper, speak into the flask, and politely request the color to change to yellow. After 8–10 students have tried, the color of the solution will suddenly change.

Procedure

Wear safety goggles. Caution: Ethyl alcohol is flammable. Extinguish all flames.

1. Prepare a flask according to the directions in the "Solutions" section prior to class.
2. Announce to the class that this reaction can be activated by just the right voice.
3. As you carry the flask through the classroom, each student should remove the stopper, speak to the solution, and stopper the flask. Give it a quick swirl each time.

Reactions

1. Eventually, CO_2 from the students' breath will produce enough acid in the solution to cause the color of the indicator to change:

$$CO_2 \text{ (g)} + H_2O \text{ (l)} \rightarrow H_2CO_3 \text{ (aq)} \rightleftarrows H^+ \text{ (aq)} + HCO_3^- \text{ (aq)}$$

2. CO_2 also reacts with NaOH. This reaction produces less basic Na_2CO_3:

$$2NaOH \text{ (aq)} + CO_2 \text{ (g)} \rightarrow Na_2CO_3 + H_2O$$

Solutions

Either of these solutions will work well:

1. Place about 250 mL of 95% ethyl alcohol in a 500-mL Florence flask. Add 5–6 drops of thymolphthalein indicator to the alcohol. Add just enough dilute NaOH to produce a blue color. Stopper the flask until it is used.
2. Prepare the solution as directed in step 1, except use 1–2 drops of phenol red in 250 mL of water. Add 1 drop of a 1 M NaOH solution to produce a red solution. Phenol red is red at pH 8.5 and turns yellow at pH 6.8.

Teaching Tips

NOTES

1. Swirling speeds up the reaction. If you have a large class, omit this step.
2. Be sure to protect the solution from CO_2 in the air prior to the demonstration.

QUESTIONS FOR STUDENTS

1. What chemistry is occurring when you speak into the flask?
2. Why does the color not change with the first student?
3. What must you consider when choosing an indicator for this demonstration?
4. Why is a drop of NaOH added prior to the demonstration?

37.   Dehydrating Action of Sulfuric Acid

A blue crystal of copper sulfate pentahydrate turns white when placed in a test tube containing concentrated sulfuric acid.

Procedure

Wear safety goggles, disposable plastic gloves, and a face shield.

1. Fill a large test tube half full of concentrated sulfuric acid. Caution: Sulfuric acid is extremely corrosive and must be handled with care.
2. Carefully add a large crystal of copper sulfate pentahydrate to the test tube.
3. Observe over several minutes that the blue of the hydrated copper sulfate disappears as the water is lost and the anhydrous white copper sulfate is formed.

Reactions

Sulfuric acid is a strong dehydrating agent. It adds water to form several known hydrates. When the water of hydration is removed from the blue copper sulfate pentahydrate, the white anhydrous form is left.

$$CuSO_4 \cdot 5H_2O \xrightarrow{H_2SO_4} CuSO_4 + 5H_2O$$
$$\text{blue} \qquad\qquad\qquad \text{white}$$

Teaching Tips

1. Concentrated sulfuric acid is 18 M. It is 98% sulfuric acid and 2% water.
2. Gases that do not react with sulfuric acid (e.g., O_2, N_2, CO_2, and SO_2) are dried by bubbling them through sulfuric acid.
3. The tendency of sulfuric acid to pick up water is so strong that it will remove water from compounds that contain hydrogen and oxygen in a 2:1 ratio even though water is not present in the molecular form. For example, sugar ($C_{12}H_{22}O_{11}$), wood, and paper starch are charred by sulfuric acid when water is removed. Sulfuric acid reacts in a similar manner with organic compounds in the skin to produce severe burns.
4. Sulfuric acid is added to acids and alcohols when esters are produced to remove the water formed during the reaction.
5. A similar demonstration (which must be done in the hood) involves adding concentrated sulfuric acid to sugar [$C_{12}(H_2O)_{11}$] in a beaker. The water in the carbohydrate sugar is removed, leaving the black, charred carbon residue. See Demonstration 38.
6. Dispose of the used sulfuric acid by adding it carefully to a large quantity of water, stirring constantly. Flush the dilute acid down the drain with copious amounts of water.
7. Sulfuric acid is one of the most important chemicals used in industry. More than 40 million tons is produced annually in the United States. It costs only about 10 cents per pound when it is produced in bulk.

QUESTIONS FOR STUDENTS

1. Explain what happens chemically when sulfuric acid acts as a dehydrating agent.
2. Show the reaction that produced anhydrous copper sulfate from the pentahydrate.
3. Will sulfuric acid also dehydrate other hydrated compounds?

38.   Dehydration of Sucrose

Sulfuric acid is poured on sugar in a beaker. In a few minutes, a large carbon snake is produced along with a puff of steam and smoke.

Procedure

Wear safety goggles, gloves, and a face shield. This demonstration must be done in a hood.

1. Fill a small beaker about one-third with table sugar (sucrose).
2. Carefully add 5–10 mL of concentrated sulfuric acid. Caution: Do not stir.
3. Observe the reactions from a safe distance of 3–5 ft.

Reactions

1. The acid will dehydrate the sugar and leave only the carbon.
2. Sulfur dioxide and acid droplets are in the fumes from the reaction.
3. The overall reaction is

$$C_{12}H_{22}O_{11} \text{ (s)} + 11H_2SO_4 \rightarrow 12C \text{ (s)} + 11H_2SO_4 \cdot 11H_2O \text{ (g)}$$

Teaching Tips

NOTES

1. Add just enough acid to wet the top half of the sugar.
2. Sucrose is $C_{12}H_{22}O_{11}$.
3. This reaction will char the beaker; therefore, use one that can be discarded after the reaction.
4. Notice that the sugar begins to char prior to the reaction.
5. Handle the carbon snake with tongs. It contains sulfuric acid.

QUESTIONS FOR STUDENTS

1. Write the equation for this reaction.
2. Would this demonstration work with another sugar, say glucose ($C_6H_{12}O_6$)?
3. Why should you not handle the carbon mass?

Energy Changes

39.   Endothermic Reaction: Ammonium Nitrate

A small amount of ammonium nitrate is added to 100 mL of water. As the solid dissolves, the temperature drops significantly.

Procedure

Wear safety goggles.

1. Place approximately 100 mL of water in a large beaker and record the temperature.
2. Quickly dump 10–15 g of ammonium nitrate into the water.
3. Note the change in temperature as the solid dissolves.

Reaction

The dissolving of ammonium nitrate in water is endothermic. The temperature should decrease 6 to 9 °C:

$$\text{heat} + NH_4NO_3 \text{ (s)} + H_2O \text{ (l)} \longrightarrow NH_4^+ \text{ (aq)} + NO_3^- \text{ (aq)}$$

Teaching Tips

NOTES

1. Theoretically, 600–900 calories are absorbed per 100 mL of water in this reaction.
2. You will get a more accurate reading by using styrofoam cups. Destroy the cups at the end of the demonstration.
3. Should you wish to calculate the heat change, 100 mL is a convenient volume.
4. One use for such a reaction is in emergency cold packs.

QUESTIONS FOR STUDENTS

1. Can you think of any practical use for such a reaction?
2. Why does the temperature drop?
3. Will the temperature drop more if twice as much solid is added?

40.   Endothermic Reaction: Two Solids

Two solids are placed in a beaker and stirred. The beaker is placed in a pool of water on a wooden block. In a few seconds, the beaker gets so cold that it freezes to the block.

Procedure

Wear safety goggles and disposable gloves. Barium hydroxide is toxic.

1. Put approximately 20 g of barium hydroxide crystals [$Ba(OH)_2 \cdot 8H_2O$] in a 50-mL beaker.
2. Add 10 g of ammonium thiocyanate to the beaker.
3. Stir the two solids together with a wooden splint.
4. Place the beaker on a small, wooden block with a small pool of water between the beaker and the block.
5. After a few minutes, the beaker will freeze to the block.
6. Dispose of the barium waste according to directions in Appendix 5.

Reaction

$$\text{heat} + Ba(OH)_2 \cdot 8H_2O \text{ (s)} + 2NH_4SCN \text{ (s)} \longrightarrow Ba(SCN)_2 \text{ (s)} + 2NH_3 \text{ (g)} + 10H_2O \text{ (l)}$$

Teaching Tips

NOTES

1. You must use barium hydroxide crystals [$Ba(OH)_2 \cdot 8H_2O$].
2. Ammonium chloride (approximately 7 g) or ammonium nitrate (10 g) may replace ammonium thiocyanate.
3. This demonstration is an excellent way to show that heats of reaction can occur without the presence of a solution.
4. Try varying the amounts of chemicals and the size of the beaker for maximum effect.

QUESTIONS FOR STUDENTS

1. What is water of hydration? Is it important in this reaction?
2. Why would heat be absorbed when water molecules are removed from hydrated barium hydroxide?
3. How does this reaction occur when the compounds are not in solution?
4. Does the size of the particles of the two solids matter?

41.   Exothermic Reaction: Calcium Chloride

Solid calcium chloride is added to water, and the temperature of the water increases.

Procedure

Wear safety goggles.

1. Note the temperature of a beaker containing 100 mL of water and quickly dump 10–15 g of calcium chloride into the beaker.
2. Record the increase in temperature.

Reaction

1. The exothermic heat of solution for calcium chloride is 117 calories per 100 mL of water:

$$CaCl_2 \text{ (s)} + H_2O \text{ (l)} \longrightarrow Ca^{2+|D} \text{ (aq)} + 2Cl^- \text{ (aq)} + heat$$

2. The temperature should increase about 12 °C.

Teaching Tips

NOTES

1. Use a styrofoam cup for a more accurate measurement of temperature change. Destroy the cups after the demonstration.
2. Barium oxide can be used instead of calcium chloride, but it is toxic.

QUESTIONS FOR STUDENTS

1. Write the chemical equation for this reaction.
2. Why does the temperature increase?
3. Why do some substances absorb heat while others liberate heat when added to water?
4. What is heat of solution?

42.   Exothermic Reaction: Sodium Sulfite and Bleach

Two solutions are mixed, and a significant increase in temperature results.

Procedure

Wear safety goggles.

1. Place 50 mL of laundry bleach in a 250-mL beaker. Record the temperature.
2. Add 50 mL of sodium sulfite solution.
3. Note the increase in temperature.

Reaction

This reaction is highly exothermic; the temperature should increase approximately 20 °C.

$$SO_3^{2-} \text{ (aq)} + 2OCl^- \text{ (aq)} \rightarrow SO_4^{2-} \text{ (aq)} + 2Cl^- \text{ (aq)} + heat$$

Solutions

1. The sodium sulfite solution is 0.5 M: 6.3 g of Na_2SO_3 per 100 mL of solution.
2. Laundry bleach is a 5.25% solution of sodium hypochlorite, NaOCl. Do not use bleaches that contain perborate.

Teaching Tips

NOTE

Use a styrofoam cup for more accurate measurement of temperature change. Destroy the cup after the demonstration.

QUESTIONS FOR STUDENTS

1. Write the chemical equation for this reaction.
2. Would the initial temperature of the solutions affect the final change in temperature?
3. Would changing the concentration of one of the solutions affect the final change in temperature?
4. Is this a redox reaction? If so, what is reduced and what is oxidized?
5. Could this demonstration be used to determine the purity of laundry bleach?

43.   Chemiluminescence: The Firefly Reaction

Two clear solutions are mixed in a darkened room. A luminescent blue color is produced that lasts for several seconds.

Procedure

Wear safety goggles and disposable gloves.

1. Place 100 mL of luminol in a flask.
2. Darken the room.
3. Add 100 mL of bleach solution to the flask.
4. Observe the reaction.

Reaction

In the presence of an oxidizing agent (bleach), luminol is converted to an excited-state product. This product decays to the ground state with the emission of light.

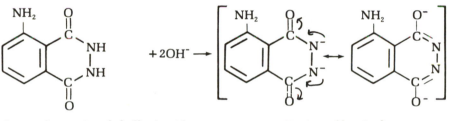

luminol (3-aminophthalhydrazide)   dianion of luminol

luminol solution

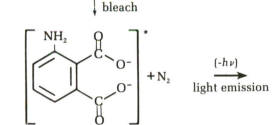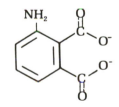

excited state of 3-aminophthalate ion   ground state of 3-aminophthalate ion

Solutions

1. Luminol solution: Dissolve 0.23 g of luminol in 500 mL of 0.1 M NaOH. Wear disposable gloves.
2. Bleach: Dilute laundry bleach (e.g., Clorox) 1:10 with water.

Teaching Tips

NOTES

1. Luminol is 5-amino-2,3-dihydro-1,4-phthalazinedione.
2. Bleach is an oxidizing agent. Oxidation occurs without the production of heat.

3. You can also use 3% hydrogen peroxide as an oxidizing agent. See Demonstration 44.

4. You can make this demonstration more dramatic by pouring the two solutions simultaneously through a funnel into a long coil of glass or noncorrosive (e.g., Tygon) tubing.

5. You can produce a dramatic effect in a darkened room by placing small pieces of filter paper on the surface of a solution made by adding 2.5 mL of 3% H_2O_2, 1-2 pellets of NaOH, and a pinch of luminol in 100 mL of water. Carefully shake a few crystals of potassium ferricyanide on each piece of paper. Drop a few crystals through the solution.

QUESTIONS FOR STUDENTS

1. Is this an oxidation reaction even though heat was not produced?
2. What does the bleach do?
3. How can you explain the production of the blue color?
4. Why is the firefly's light yellow?
5. Is biology really chemistry?

44. Chemiluminescence: Two Methods

Method 1

Luminol is added to a jar containing a solution. A brilliant blue is produced that lasts for several minutes.

Procedure

Wear safety goggles and plastic gloves.

1. Place 70 g of KOH in a quart jar with a screw cap.
2. Add 60 mL of dimethyl sulfoxide (DMSO).
3. Pass a stream of oxygen gas into the bottle for a few seconds.
4. Cap the jar.
5. When you are ready for the demonstration, add 0.1 g of luminol, shake the jar gently for a few minutes, and pass it around the class.

Reactions

See Demonstration 43 for reactions.

Solutions

1. DMSO is dimethyl sulfoxide. You can buy this chemical at stores where farm supplies are sold. Caution: DMSO penetrates the skin rapidly and carries with it any toxic substances that may be on the skin's surface. Wear plastic gloves when you use this solvent.
2. If you do not have a tank of oxygen, you may be able to borrow one from the shop or maintenance department.

Method 2

Two solutions are mixed; a luminescence is produced for several seconds.

Procedure

1. Mix equal volumes of solution A and solution B.
2. Observe the chemiluminescence.

Reactions

See Demonstration 43 for reactions.

Solutions

1. Solution A: Dissolve 4 g of sodium carbonate in approximately 500 mL of water. Add a small amount of luminol (approximately 0.2 g) and stir until it dissolves. Add 0.5 g of ammonium bicarbonate monohydrate. Add 0.5 g of copper sulfate. Add 25 g of sodium bicarbonate. When everything has dissolved, dilute to 1 L.
2. Solution B: Dilute 50 mL of 3% H_2O_2 (drugstore variety) to 1 L.

Teaching Tips

NOTES

1. These reactions use oxidizers other than bleach.
2. These variations produce a more brilliant luminescence that lasts longer than that oxidized by bleach (see Demonstration 43).

QUESTIONS FOR STUDENTS

1. What is the role of hydrogen peroxide?
2. How was the blue color produced?
3. Why was the DMSO solution saturated with oxygen?

45.   Energy and the Rubber Band

Students are asked to hold a rubber band to their foreheads. When the rubber band is suddenly stretched, a noticeable amount of heat is produced and can be felt on the skin. When the rubber band is relaxed, a slightly cool sensation is felt. Following up on this, heat is applied to a rubber band stretched by a weight. The rubber band contracts.

Procedure

Wear safety goggles.

1. Hold a rubber band tightly across your forehead, cheek, or lips. Stretch the rubber band and notice that heat is produced (an exothermic process). Hold it for 30 s, then relax the rubber band and note that it feels cool (an endothermic process).
2. Hook a similar rubber band over a clamp on a ring stand. Hook one end of a bent paper clip on the other end of the rubber band and add large lead fishing weights to stretch the rubber band. Place a metric ruler behind the band so that the length of the stretched band can be measured. Record this length.
3. Using a hair dryer set for maximum heat, warm the rubber band.
4. Observe that the band contracts. Measure the length of the warmed rubber band.

Reactions

This is a fascinating demonstration because it produces results that might not be expected. There is a more orderly arrangement of molecules in a stretched rubber band than in a relaxed rubber band. When the band is stretched, entropy decreases and heat is produced in an exothermic process. When the orderly, stretched rubber band is heated, the process is endothermic but spontaneous, and entropy increases.

Teaching Tips

1. Do not use a burner or open flame that might burn the rubber.
2. Try varying the amount of lead weights to get the best results. The rubber band should be stretched to its maximum extent for best results.
3. Entropy, S, is a measure of disorder in a system. The change in entropy of a system, ΔS, is a function of both heat flowing in and out of the system and of the temperature. Disorder tends to increase spontaneously. As the rubber band relaxes (spontaneous), the entropy (disorder) must increase.
4. Encourage students to learn more about entropy by referring them to any general chemistry textbook.
5. We can tell that the molecules have become more ordered because S must be negative; entropy (disorder) has decreased. $\Delta H = \Delta G + T\Delta S$. If ΔH is negative (exothermic), ΔG is positive (nonspontaneous), T (absolute temperature) is positive, then ΔS must be negative (more ordered) in this system.

QUESTIONS FOR STUDENTS

1. If the rubber band gets warm when stretched, why does that mean the molecules have become more ordered?
2. How do you know the relaxing of the rubber band is spontaneous?
3. Explain why relaxing of the rubber band is endothermic.

Equilibrium

46.   Equilibrium and LeChatelier's Principle

This demonstration involves the equilibrium reaction

$$Fe^{3+}\ (aq) + SCN^-\ (aq) \rightleftharpoons FeSCN^{2+}\ (aq)$$

By using a Petri dish on an overhead projector, you can show the color changes resulting from changes in the concentration of reactants.

Procedure

Wear safety goggles.

1. Place a Petri dish on a clear plastic sheet on an overhead projector.
2. Cover the bottom of the Petri dish with the KSCN solution. To indicate the ions present, write K^+ (aq) and SCN^- (aq) on the plastic sheet.
3. Add 2–3 drops of $Fe(NO_3)_3$ solution. Note the color change. Write Fe^{3+} (aq) and NO_3^- (aq) to show the ions that were added.
4. Because the color change indicates the formation of a new species, show students that this reaction must proceed as follows:

$$Fe^{3+}\ (aq) + SCN^-\ (aq) \rightleftharpoons FeSCN^{2+}\ (aq)$$

5. Add a small crystal of KSCN to the dish. Do not stir. Notice the formation of a darker color from $FeSCN^{2+}$ (aq). The darker color represents a shift of equilibrium to the right.
6. Add a drop of $Fe(NO_3)_3$ solution. Notice that the color again intensifies, and a shift to the right is indicated.
7. Explain to the students that you can remove some Fe^{3+} by complexing it with Na_2HPO_4. Add a small crystal of Na_2HPO_4; note the immediate clearing of color. A shift of equilibrium to the left is thus indicated.

Reactions

1. The additional SCN^- (aq) from KSCN increases the concentration of reactants and causes a shift of equilibrium toward the products.
2. The additional Fe^{3+} (aq) from $Fe(NO_3)_3$ increases the concentration of reactants and causes a shift of equilibrium toward the right.
3. Adding Na_2HPO_4 reduces the concentration of Fe^{3+} (aq) by forming the colorless complex, $FeHPO_4^+$ (aq). This reaction causes a shift of equilibrium to the left and the formation of a lighter color.

Solutions

1. The KSCN solution is 0.002 M: Dissolve 0.19 g of KSCN per liter.
2. The $Fe(NO_3)_3$ solution is 0.2 M: Dissolve 8 g of $Fe(NO_3)_3 \cdot 9H_2O$ per 100 mL of water. Store this solution in a dropper bottle; very little is used.

Teaching Tips

NOTES

1. You can do this demonstration in beakers or project it in a Petri dish.
2. Have students make predictions before you perform each part of the demonstration.
3. Do not confuse the poisonous cyanide ion, CN^-, with the nontoxic thiocyanate ion, SCN^-.
4. You can restore an equilibrium system after the last part of the demonstration by adding Na_2HPO_4.

QUESTIONS FOR STUDENTS

1. Why are K^+ (aq) and NO_3^- (aq) ions not included in the equilibrium reaction?
2. When a crystal of KSCN was added, the solution became darker as a result of the formation of more $FeSCN^{2+}$ (aq). How could more $FeSCN^{2+}$ (aq) form if no additional Fe^{3+} (aq) was added?
3. What would you suggest be done to restore an equilibrium system after the last part of the demonstration?

47. Effect of Temperature Change on Equilibrium: Cobalt Complex

An equilibrium system involving the dehydrated–hydrated cobalt complex is produced. When this system is heated, a color change from pink to blue indicates a shift of equilibrium to the right. When the solution is cooled, the color change from blue to pink indicates a shift to the left.

Procedure

Wear safety goggles and disposable plastic gloves. $CoCl_2$ is toxic.

1. Place 100 mL of $CoCl_2$ solution in a 250-mL beaker.
2. Add concentrated HCl until the solution changes from pink to blue. Use a face shield and gloves when you use concentrated HCl.
3. Divide the solution into three smaller beakers and treat them as follows:
   a. Place one beaker on a hot plate.
   b. Place one beaker in an ice bath.
   c. Leave one beaker at room temperature as a standard.
4. After a few minutes, show that the heated sample has turned a darker blue and that the cooled sample has turned a light pink.

Reactions

1. This reaction involves the following equilibrium:

$$\text{heat} + [Co(H_2O)_6]^{2+} \text{ (aq)} + 4Cl^- \text{ (aq)} \rightleftharpoons [CoCl_4]^{2-} \text{ (aq)} + 6H_2O$$
$$\text{pink} \qquad\qquad\qquad\qquad\qquad\qquad\qquad \text{blue}$$

2. Addition of heat causes a shift of equilibrium toward products, the blue solution.
3. Cooling causes a shift of equilibrium to the left, the pink hydrated complex.

Solutions

1. The $CoCl_2$ solution is 0.4 M: Dissolve 5.2 g per 100 mL of water.
2. The HCl solution is concentrated.

Teaching Tips

NOTES

1. As indicated in the equation, you may have to add quite a bit of HCl to get the formation of the blue complex.
2. The blue color is due to the tetrachlorocobalt(II) complex, and the pink color is due to the hexaaquacobalt(II) complex.
3. For an interesting variation, heat 150 mL of $CoCl_2$ solution until it turns blue. Fill a large test tube with this solution and immerse it halfway into a beaker that contains crushed ice and salt. The bottom part of the test tube will turn pink.

QUESTIONS FOR STUDENTS

1. Write an equation for the equilibrium system.
2. Why was it necessary to add HCl to establish equilibrium?
3. How does heating shift the equilibrium?
4. What do you think will happen to the equilibrium system if you add water?

48.   Effect of Concentration on Equilibrium: Cobalt Complex

Changing the concentration of reactants results in a shift of equilibrium between pink and blue complex ions of cobalt.

Procedure

Wear safety goggles and disposable plastic gloves. $CoCl_2$ is toxic.

1. Shift of equilibrium to the right: Place 20 mL of cobalt(II) chloride solution in a small beaker. Slowly add 40 mL of concentrated HCl. Note the formation of blue color. Use a face shield and gloves when you use concentrated HCl.
2. Shift of equilibrium to the left: Use half of the solution from step 1; save the other half. Add 20 mL of distilled water. Note the color change to pink.
3. Shift of equilibrium to the left: To the remaining solution from step 1, add silver nitrate solution a drop at a time until a precipitate forms. Note the formation of a pink color.
4. Dispose of the solutions according to the directions in Appendix 5.

Reactions

1. This demonstration involves the following equilibrium:

$$[Co(H_2O)_6]^{2+} \text{ (aq)} + 4Cl^- \text{ (aq)} \rightleftarrows [CoCl_4]^{2-} \text{ (aq)} + 6H_2O \text{ (l)}$$

    pink   blue

2. Excess Cl^- causes the formation of more blue tetrachlorocobalt(II) complex.
3. Excess water shifts the equilibrium to the left and forms more pink hexaaqua-cobalt(II) complex.
4. Silver nitrate removes Cl^-, and the precipitate silver chloride is formed. This precipitation causes the equilibrium to shift to the left, and more pink complex is formed.

Solutions

1. The cobalt(II) chloride solution is 0.2 M: 26 g of $CoCl_2$ per liter of water.
2. The hydrochloric acid is a concentrated solution.
3. The silver nitrate solution is 0.1 M: 1.7 g of $AgNO_3$ per 100 mL of distilled water (distilled *only*). Caution: Use gloves with silver nitrate.

Teaching Tips

NOTES

1. This demonstration projects well. Use Petri dishes on an overhead projector.
2. Be sure to use concentrated HCl in step 1. In addition to adding Cl^-, HCl has a dehydrating effect.
3. Have students make predictions of the effect on equilibrium before each part of the demonstration.

QUESTIONS FOR STUDENTS

1. What is the effect of adding a common ion to a system in equilibrium?
2. Explain what happened in each part of the demonstration.
3. What might be another way to shift equilibrium to the right?
4. Blue colors are usually associated with hydrated compounds. Why does the hydrated cobalt complex have a pink color?

49.   Effects of Concentration and Temperature on Equilibrium: Copper Complex

An equilibrium system involving the blue Cu^{2+} and green copper complex is prepared. Changes are made in the concentration of reactants and the temperature to shift the equilibrium.

Procedure

Wear safety goggles.

1. Place about 150 mL of $CuSO_4$ solution in a flask. Note the blue color.
2. Add 50 mL of KBr solution. Note the color change from blue to green.
3. Divide the solution into two equal parts for demonstrating the effect of concentration and the effect of temperature.
4. Show the effect of changing concentration: Add a little solid Na_2SO_4 and note the color change from green to blue. Add concentrated HCl to the solution and note the color change from blue to green. Use a face shield and gloves when you handle concentrated HCl.
5. Show the effect of changing temperature: Place the beaker from step 4 in an ice bath and note the color change from green to blue. Remove the beaker from the ice bath, heat it, and note the color change from blue to green.

Reactions

1. The equilibrium reaction is as follows:

$$\text{heat} + \underset{\text{blue}}{CuSO_4} \ (aq) + 4KBr \ (aq) \rightleftharpoons \underset{\text{green}}{K_2[CuBr_4]} \ (aq) + K_2SO_4 \ (aq)$$

or, simply:

$$Cu^{2+} \ (aq) + 4Br^- \ (aq) \rightleftharpoons CuBr_4^{2-} \ (aq)$$

2. Adding KBr shifts the equilibrium to the right and forms more green complex.
3. Adding Na_2SO_4 forms a basic solution and shifts the equilibrium to the left, forming more blue Cu^{2+}; adding H^+ shifts equilibrium to the right (toward green).
4. Heating forces the equilibrium to the right; cooling pushes the reaction to the left.

Solutions

1. The KBr solution is saturated: 50 g in 50 mL of hot water. Solid KBr can also be used.
2. The $CuSO_4$ solution is 0.2 M: 50 g of $CuSO_4 \cdot 5H_2O$ per liter of water.
3. The HCl is concentrated.

Teaching Tips

NOTES

1. These colors project well and can be demonstrated with Petri dishes and an overhead projector.
2. Do not get involved, at this point, with the structure of the copper complex.

QUESTIONS FOR STUDENTS

1. Is this an endothermic or exothermic reaction?
2. How does heating shift the equilibrium?
3. How does adding KBr shift the equilibrium?
4. How does adding Na_2SO_4 cause the equilibrium to shift?

50. Equilibrium in the Gas Phase

A reddish-brown gas is prepared and placed in two small test tubes. When one test tube is placed in boiling water, the color of the gas changes to a deep brown color. When the other test tube is placed in an ice bath, the gas becomes almost colorless. When both tubes are allowed to reach room temperature, the gas in both again becomes reddish-brown.

Procedure

Wear safety goggles.

Preparing the gas test tubes

1. Prepare a gas generator by attaching a rubber tube to a small side-arm flask.
2. In a hood, add 10 mL of concentrated nitric acid to the flask. Use a face shield and gloves when you handle concentrated nitric acid.
3. Drop in a copper penny.
4. A deep-red gas, NO_2, will immediately form. Allow enough of the gas to form to displace the air in the flask and two test tubes. Colorless NO forms first. Colorless bubbles rise to the surface, where they mix with O_2 in the air and immediately form NO_2.
5. Fill the two test tubes with the brown gas. Stopper the tubes.
6. Stop the reaction in the flask by filling the flask with water. Notice the blue color of the solution and the small size of the penny.

Demonstrating equilibrium

1. Point out the color of the gas in the test tube. Put the equilibrium reaction on the board.
2. Place one test tube in a beaker of boiling water. Notice the change in color.
3. Place the other test tube in an ice bath. Notice the formation of a colorless gas.
4. Remove both test tubes and allow them to come to room temperature. Note the restoration of the brown color in both test tubes.

Reactions

1. The equations for the production of the gas follow:

$$3Cu + 8H^+ + 2NO_3^- \longrightarrow 3Cu^{2+} + 2NO\ (g) + 4H_2O$$

$$2NO\ (g) + O_2\ (g) \longrightarrow 2NO_2\ (g)$$

2. The equilibrium mixture in the test tubes consists of NO_2 and N_2O_4. They react according to the equation:

$$\underset{\text{red}}{2NO_2\ (g)} \rightleftharpoons \underset{\text{colorless}}{N_2O_4\ (g)}$$

3. When the equilibrium mixture is heated, the equilibrium shifts toward the formation of brown NO_2.
4. When the mixture is cooled, the equilibrium shifts toward the formation of more colorless N_2O_4.

Teaching Tips

NOTES

1. The preparation of the gas test tubes must be done in a hood because NO_2 is toxic. It has been estimated that two pennies could form enough poisonous NO_2 to produce a toxic level in a typical laboratory.
2. Fat, short, clear plastic test tubes work best. They can be used for several weeks if they are stoppered tightly.

QUESTIONS FOR STUDENTS

1. Write the chemical equation for the production of the gas.
2. Why is the solution formed in the reaction vessel blue?
3. Could the chemical change in the test tubes be due to an increase in pressure rather than an increase in temperature?
4. Devise an experiment to prove or disprove your hypothesis.
5. Is this an exothermic or endothermic reaction?

51. Equilibrium: Temperature and the Ammonium Hydroxide–Ammonia System

A large test tube containing a pink solution is heated and the color disappears. The test tube is cooled in tap water and the pink color returns.

Procedure

Wear safety goggles.

1. Place about 400 mL of water in a large beaker.
2. Add a few drops of phenolphthalein indicator.
3. Add 1 drop of concentrated ammonia. The solution should be pink, but not dark pink.
4. Fill a large test tube half full of this solution.
5. Heat the test tube over a burner. Note that the pink color fades and disappears.
6. Let the test tube cool for a few seconds, then place it in a stream of cold water. Notice that the pink color returns.

Reactions

1. The reaction is basically an equilibrium between the ammonium hydroxide and the nonionized ammonia:

$$NH_4^+ \text{ (aq)} + OH^- \text{ (aq)} \rightleftharpoons NH_3 \text{ (aq)} + H_2O \text{ (l)}$$

2. Heat shifts the equilibrium and reduces the ammonium hydroxide so that it no longer reacts with the indicator.

Teaching Tips

NOTES

1. If you add too much ammonia, the pink color will persist when the test tube is heated.
2. You can do this demonstration in a larger container, but it will take longer to cool and restore the color.

QUESTIONS FOR STUDENTS

1. Write the reaction for this equilibrium.
2. Is the reaction endothermic or exothermic?
3. Why does the demonstration not work well if too much ammonia is added?

52.   Effect of Pressure on Equilibrium

A side-arm flask containing a colorless solution is connected to an aspirator on a water faucet. When the water faucet is turned on, the solution begins to bubble and later turns pink.

Procedure

Wear safety goggles.

1. Fit a 250-mL side-arm flask with a tight-fitting rubber stopper and vacuum tubing connected to the aspirator on a laboratory water faucet.
2. Add 150 mL of $NaHCO_3$ solution to the flask.
3. Add 3–4 drops of phenolphthalein indicator. If the solution is pink, add dilute HCl a drop at a time until the color just disappears.
4. Stopper the flask and turn on the water.
5. Notice the formation of a gas and the eventual color change to pink.

Reactions

1. The equilibrium system is represented by the following scheme:

$$HCO_3^- \text{ (aq)} \rightleftharpoons OH^- \text{ (aq)} + CO_2 \text{ (g)}$$

2. Decreasing pressure forces the equilibrium to the right. It also causes the release of CO_2 and causes the solution to become more alkaline.
3. The alkaline solution causes the indicator to become pink.

Solution

The sodium bicarbonate solution is saturated: Dissolve 32 g of $NaHCO_3$ in 200 mL of water.

Teaching Tips

NOTES

1. Students should know that phenolphthalein is colorless in acid solution and pink in alkaline solution.
2. In addition to reducing the pressure, the vacuum is removing a product (CO_2).

QUESTIONS FOR STUDENTS

1. Write the equilibrium reaction for this system.
2. Why does the solution become pink?
3. Which is more significant in this change, reducing pressure or removing CO_2?

53.   Effect of Hydrolysis on Equilibrium

A beaker contains a clear solution. When water is added to the solution, a dense, white precipitate forms. When HCl is added, the precipitate dissolves and the solution becomes clear again.

Procedure

Wear safety goggles and disposable plastic gloves. $SbCl_3$ and $BiCl_3$ are toxic.

1. Place about 250 mL of water in a beaker.
2. Add a small amount (one or two small lumps) of antimony trichloride ($SbCl_3$). Note the cloudy solution.
3. Add concentrated HCl until the solution just clears. Use a face shield and gloves when you handle concentrated HCl.
4. Add water until a precipitate again appears and the solution becomes cloudy.
5. Dispose of the solid waste according to the directions given in Appendix 5.

Reactions

1. The equilibrium system is as follows:

$$\underset{\text{clear}}{SbCl_3 \text{ (s)}} + H_2O \text{ (l)} \rightleftharpoons \underset{\text{cloudy}}{SbOCl \text{ (s)}} + 2HCl \text{ (aq)}$$

2. Adding HCl shifts the equilibrium to the left and results in clear antimony trichloride solution.
3. Adding water shifts the equilibrium to the right and results in white antimony(III) oxychloride, SbOCl.

Teaching Tips

NOTES

1. This reaction projects well. Use Petri dishes and an overhead projector.
2. Use the smallest amount of $SbCl_3$ possible to give a cloudy solution. Otherwise, a large volume of HCl will be needed to clear the solution.
3. You can also do this demonstration with bismuth chloride; bismuthyl chloride is formed as a hydrolysis product:

$$\underset{\text{clear}}{BiCl_3 \text{ (s)}} + H_2O \text{ (l)} \rightleftharpoons \underset{\text{white}}{BiOCl \text{ (s)}} + 2HCl \text{ (aq)}$$

QUESTIONS FOR STUDENTS

1. What is hydrolysis?
2. How does antimony trichloride change when water is added?
3. Would addition of a chloride (e.g., ammonium chloride) rather than HCl cause the equilibrium to shift?
4. Can you think of other things to do to cause a shift in equilibrium?

54.   Solubility Product: Effect of Concentration

A saturated solution of lead bromide is prepared. When a solution of lead nitrate is added, no precipitate forms. However, when a solution of sodium bromide is added, extensive precipitation occurs.

Procedure

Wear safety goggles and disposable gloves. Lead compounds are toxic.

1. Place about 150 mL of lead bromide solution in two beakers.
2. In one beaker add 10–15 mL of lead nitrate solution. Notice that no precipitate forms.
3. In the other beaker, add the same amount of sodium bromide solution. Notice that precipitation is heavy.
4. Dispose of the lead compounds according to the directions in Appendix 5.

Reaction

$$PbBr_2 \ (s) \rightleftharpoons Pb^{2+} \ (aq) + 2Br^- \ (aq)$$

This reaction clearly shows that the concentration of Br^- affects the solubility product expression for lead bromide to a greater degree than does the concentration of lead.

$$K_{sp} = [Pb^{2+}][Br^-]^2 = 6.3 \times 10^{-6}$$

Solutions

1. The lead nitrate solution is 1.0 M: 331 g of $Pb(NO_3)_2$ per liter of solution.
2. The sodium bromide solution is 1.0 M: 103 g of NaBr per liter of solution.
3. Prepare the lead bromide as follows: Add 100 mL of 1.0 M lead nitrate solution to 200 mL of 1.0 M sodium bromide solution. Filter the solution to remove the precipitated lead bromide. You may need to use suction filtration.
4. Prepare a saturated solution of lead bromide: 10 g of $PbBr_2$ in 200 mL of water.

Teaching Tips

NOTES

1. Use the lead bromide solution sparingly.
2. Lead bromide is only slightly soluble. Use warm water.
3. This demonstration projects well. Use Petri dishes and an overhead projector.

QUESTIONS FOR STUDENTS

1. What is a solubility product constant?
2. Write the expression for lead bromide.
3. Why does additional Pb^{2+} cause no precipitation of lead bromide, whereas additional Br^- does?
4. How is equilibrium involved in this reaction?

55. The Common Ion Effect: First Demonstration

Two Erlenmeyer flasks contain acetic acid. Solid sodium acetate is added to one bottle. Magnesium metal is added to each flask, and balloons are attached. The balloon on the flask containing the sodium acetate requires longer to inflate than does the balloon on the other flask.

Procedure

Wear safety goggles.

1. Fill two Erlenmeyer flasks one-fourth full of acetic acid.
2. Add approximately 15 g of sodium acetate to one of the flasks.
3. Add about 3 g of magnesium powder to each flask.
4. Fit the flasks with balloons.
5. Observe and compare the reactions in the two flasks.

Reactions

1. Hydrogen is produced in each flask as a result of the action of acid on magnesium metal.

$$2CH_3COOH \text{ (aq)} + Mg \text{ (s)} \rightarrow H_2 \text{ (g)} + Mg(CH_3COO)_2 \text{ (aq)}$$

2. The concentration of hydrogen from the acid has been reduced by the presence of additional acetate ion.

$$2CH_3COOH \text{ (aq)} \rightleftharpoons H_2 \text{ (g)} + \underset{\text{increased}}{2CH_3COO^- \text{ (aq)}}$$

3. The rate of the reaction is directly proportional to the concentration of hydronium ion. Therefore, the rate of hydrogen production in the buffered solution is less than the rate in the flask containing only acetic acid.
4. Both balloons eventually reach the same size. Thus, the total amount of available hydrogen was not changed in the buffered flask, only the rate of formation.

Solutions

1. The acetic acid solution is approximately 2 M. Try using vinegar (5%) diluted to half strength (see Appendix 2).
2. Use about 15 g of sodium acetate. Shake the flask to dissolve the sodium acetate in the acetic acid.

Teaching Tips

NOTES

1. Place the balloons on the flasks quickly and simultaneously. Have students help, and go through a few dry runs.
2. Secure the balloons to the flasks with tape or twist ties.

3. It may take 5–8 min for both balloons to reach the same size. Use this time to discuss the common ion effect.
4. Party balloons work best. They have wide mouths.

QUESTIONS FOR STUDENTS

1. Write the equation for the reaction in each flask.
2. What effect does the added acetate have on the rate of hydrogen formation? What effect does it have on the total amount of hydrogen formed?
3. Is this reaction an oxidation–reduction reaction? If so, what is oxidized?

56.   The Common Ion Effect: Second Demonstration

This demonstration is a variation of the reaction in Demonstration 55, but the gas produced in the reaction causes the production of a foam.

Procedure

Wear safety goggles.

1. Place a heaping teaspoon (7–8 g) of precipitated calcium carbonate in each of two 500-mL graduated cylinders.
2. Add 100 mL of solution A to one cylinder.
3. Add 100 mL of solution B to the other cylinder.
4. Note the height of foam produced in each cylinder. Mark the level with a wax pencil or with tape.

Reactions

1. Calcium carbonate reacts with acetic acid to produce CO_2 gas.

$$CaCO_3 \text{ (s)} + 2CH_3COOH \text{ (aq)} \rightarrow Ca^{2+} \text{ (aq)} + 2CH_3COO^- \text{ (aq)} + H_2O \text{ (l)} + CO_2 \text{ (g)}$$

2. The height of foam produced is proportional to the amount of gas produced, and the rate of foam production is proportional to the rate of carbon dioxide production.

Solutions

1. Solution A: 100 mL of 2–3 M acetic acid (see Appendix 2).
2. Solution B: 100 mL of 2–3 M acetic acid to which a heaping teaspoon (about 10 g) of solid sodium acetate has been added; shake the solution to dissolve the solid.

Teaching Tips

NOTES

1. The cylinder containing acetic acid with sodium acetate will produce carbon dioxide more slowly than will the other cylinder because of the additional acetate ion. Thus, rapid foam production will not occur.
2. The total amount of foam in each cylinder will be the same because the equilibrium in the buffered solution ultimately shifts to produce the same amount of carbon dioxide ion as the unbuffered solution.
3. Try varying the amounts of calcium carbonate and acetic acid to achieve best results.
4. Use laboratory-grade (precipitated) calcium carbonate, not the purified analytical-grade.
5. This demonstration offers an excellent opportunity to discuss buffering in acid–base reactions.

QUESTIONS FOR STUDENTS

1. What effect does additional acetate ion have on the production of foam?
2. Write the equation for this reaction.
3. What is the common ion effect?
4. When a potassium chloride solution is added to a clear solution of potassium perchlorate, potassium perchlorate precipitates as a milky white substance. Try it and explain the result. What is the common ion?

57. The Common Ion Effect: Ammonium Hydroxide and Ammonium Acetate

A large beaker contains a light red solution of NH^+ (aq). A small amount of solid NH_4CH_3COO is added and the solution becomes colorless.

Procedure

Wear safety goggles.

1. Add 250 mL of water and a few drops of phenolphthalein solution to a beaker.
2. Add 1 M aqueous ammonia drop by drop until the color just changes to light red.
3. Add a small amount of solid ammonium acetate or ammonium chloride.
4. Note the change in color from red to colorless.

Reactions

$$NH_3 \text{ (aq)} + H_2O \text{ (l)} \rightleftharpoons NH_4^+ \text{ (aq)} + OH^- \text{ (aq)}$$

1. Adding ammonium chloride increases the concentration of NH_4^+:

$$NH_4Cl \text{ (s)} \rightarrow NH_4^+ \text{ (aq)} + Cl^- \text{ (aq)}$$

2. As additional NH_4^+ is added, the equilibrium shifts to the left, and the concentration of OH^- is reduced to the point that the indicator turns colorless (below pH 9.5).

$$NH_3 \text{ (aq)} + H_2O \text{ (l)} \rightleftharpoons \underset{\text{increased}}{NH_4^+} \text{ (aq)} + OH^- \text{ (aq)}$$

Solutions

Aqueous ammonia, NH_3 (aq), is 1.0 M. See Appendix 2.

Teaching Tips

NOTES

1. You can add sodium acetate or sodium chloride to the solution to show that the precipitate is not due to the effect of the acetate or the chloride.
2. Try other indicators.

QUESTIONS FOR STUDENTS

1. Define the common ion effect.
2. What is the common ion in this demonstration?
3. Write the equation for this reaction.
4. What would happen to the equilibrium if NaOH is added?

Kinetics

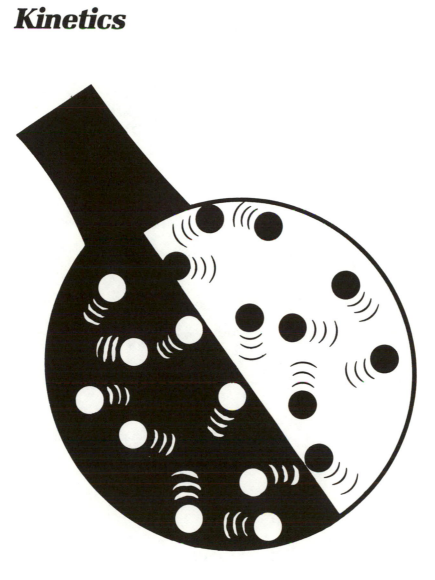

58.   Rare Earth Oxides as Catalysts

An attempt is made to burn a cube of sugar held on a toothpick. It does not burn; it only melts when heated with a match. After a second sugar cube is dipped into a pile of cigarette ashes, it can be burned when lit with a match.

Procedure

Wear safety goggles.

1. Show your students that you cannot burn a cube of table sugar held on a toothpick (or with tongs) with the heat from a match.
2. Take another cube, dab it in a pile of cigarette ashes; cover at least two sides of the cube. When the cube is heated with a match, the cube will burn.

Reaction

The rare earth oxides in the cigarette ashes act as catalysts in the combustion of sugar.

Teaching Tips

NOTES

1. Fine carbon will work as well as cigarette ashes.
2. The melting point of sucrose is 185 °C.
3. Do not leave the impression with students that cigarette ashes are normally found in the laboratory.

QUESTIONS FOR STUDENTS

1. What is a catalyst?
2. How do these oxides act as catalysts?
3. What other substances can you think of that act as catalysts?

59.   Medicine Cabinet Kinetics: How Fast Is the Fizz?

Fresh Alka Seltzer tablets are dropped simultaneously into beakers of ice water and hot water. The one dropped in hot water fizzes and reacts much faster than the one dropped in ice water.

Procedure

Wear safety goggles.

1. Place approximately 150 mL of hot water in a beaker.
2. Place the same amount of ice water in a second beaker.
3. Drop an Alka Seltzer tablet into each beaker.
4. Note the rate of reaction in each beaker.

Reactions

1. Alka Seltzer consists of a mixture of the following compounds: calcium dihydrogen phosphate, $Ca(H_2PO_4)_2$; sodium bicarbonate, $NaHCO_3$; citric acid; and aspirin.
2. Calcium dihydrogen phosphate is a source of hydrogen ion:

$$Ca(H_2PO_4)_2 \text{ (s)} \rightarrow 2H^+ \text{ (aq)} + 2(HPO_4)^{2-} \text{ (aq)} + Ca^{2+} \text{ (aq)}$$

3. Sodium bicarbonate is a source of bicarbonate ion, HCO_3^-:

$$NaHCO_3 \text{ (s)} \rightarrow Na^+ \text{ (aq)} + HCO_3^- \text{ (aq)}$$

4. H^+, HCO_3^-, and H_2O react to produce carbon dioxide and water:

$$H^+ \text{ (aq)} + HCO_3^- \text{ (aq)} + H_2O \text{ (l)} \rightarrow CO_2 \text{ (g)} + 2H_2O \text{ (l)}$$

5. The overall reaction is

$$Ca(H_2PO_4)_2 \text{ (s)} + 2NaHCO_3 \text{ (s)} \rightarrow 2CO_2 \text{ (g)} + 2H_2O \text{ (l)} + CaHPO_4 \text{ (s)} + Na_2HPO_4 \text{ (s)}$$

Teaching Tips

NOTES

1. Use fresh Alka Seltzer tablets.
2. This reaction occurs when baking powder, which also contains calcium dihydrogen phosphate and sodium bicarbonate, decomposes and causes baked products to rise.

QUESTIONS FOR STUDENTS

1. What generalization about rates of reactions can be drawn from this demonstration?
2. If the Alka Seltzer tablets are old, they do not produce much fizz. Why?
3. How would the rate of reaction in water at room temperature compare with that in cold water? Hot water?

60.   Catalytic Decomposition of Hydrogen Peroxide: Foam Production

A tremendous amount of foam shoots from a graduated cylinder when detergent and potassium iodide are added to hydrogen peroxide.

Procedure

Wear safety goggles and disposable gloves.

1. Place a 500-mL graduated cylinder in a plastic tray or a laboratory sink.
2. Pour approximately 50 mL of 30% hydrogen peroxide into the cylinder. Caution: Handle 30% H_2O_2 only with plastic gloves.
3. Add a squirt of dishwashing detergent and a drop of food color.
4. Add about ¼ of a spoon (1–2 g) of solid KI.

Reactions

1. The rapidly catalyzed decomposition of hydrogen peroxide produces oxygen gas, which forms a foam with the liquid detergent:

$$2H_2O_2 \text{ (aq)} \longrightarrow 2H_2O \text{ (l)} + O_2 \text{ (g)}$$

2. The decomposition of H_2O_2 in the presence of iodide ion occurs in two steps. The first reaction is the rate-determining reaction.

$$H_2O_2 \text{ (aq)} + I^- \text{ (aq)} \longrightarrow H_2O \text{ (l)} + OI^- \text{ (aq)}$$

$$H_2O_2 \text{ (aq)} + OI^- \text{ (aq)} \longrightarrow H_2O \text{ (l)} + O_2 \text{ (g)} + I^- \text{ (aq)}$$

Solution

Use 30% hydrogen peroxide for best results. Do not use the 3% hydrogen peroxide available in drugstores. Use extreme care with 30% H_2O_2.

Teaching Tips

NOTES

1. Be careful when using 30% hydrogen peroxide. Wear gloves and avoid contact with this solution. Always store 30% hydrogen peroxide in an approved refrigerator.
2. You can also show the decomposition of hydrogen peroxide on an overhead projector. Place a small amount of 3% hydrogen peroxide in a Petri dish on an overhead projector. Add a pinch of potassium iodide or manganese dioxide and note the evolution of oxygen gas bubbles.
3. The catalytic decomposition of hydrogen peroxide occurs when the 3% solution is placed on a wound. Catalase, an enzyme in the blood, catalyzes the reaction.
4. The brown color of the foam is evidence of iodine in the reaction.

QUESTIONS FOR STUDENTS

1. How does a catalyst work?
2. What happened to the KI?
3. How can you account for the large amount of foam produced?
4. What evidence is there that iodine is produced?

61. A Catalyst in Action

A catalyst is added to a solution. A green activated complex is formed and the reaction proceeds. When the green complex disappears, the reaction ceases.

Procedure

Wear safety goggles and disposable plastic gloves. $CoCl_2$ is toxic.

1. Place 200 mL of sodium potassium tartrate solution in a 600-mL beaker.
2. Warm the solution gently to about 70 °C on a hot plate. Use caution.
3. Add 80 mL of hydrogen peroxide.
4. Add a few crystals (about a spoon, 5–7 g) of cobalt(II) chloride.
5. A vigorous reaction will occur. Note the appearance of the green activated complex and the extensive bubbling.

Reactions

1. This reaction involves the oxidation of tartaric acid, $HO_2CCH(OH)CH(OH)CO_2H$, by hydrogen peroxide in the presence of a cobalt(II) chloride catalyst.
2. The green color is due to the formation of a cobalt–tartrate activated complex.
3. Note that the original catalyst, cobalt(II) chloride, is pink when in aqueous solution. As the tartrate is oxidized, the activated complex is broken down to the original catalyst, and the pink color returns.
4. Oxygen and carbon dioxide gases are produced. Oxalic acid, HO_2CCO_2H, is probably produced also.

Solutions

1. Sodium potassium tartrate: 12 g of $KNaC_4H_4O_6 \cdot 4H_2O$ per 200 mL of distilled water.
2. Hydrogen peroxide: Use 6% H_2O_2 for this reaction, *not* the common 3% variety available in drugstores. You can obtain 6% H_2O_2 at drugstores (Clairoxide is one brand), or you can dilute 30% hydrogen peroxide by adding 200 mL of 30% H_2O_2 per liter of solution. Be careful with 30% H_2O_2. Wear gloves when you handle it.

Teaching Tips

NOTES

1. This demonstration is one of the few that allows us to observe the formation and action of an activated complex.
2. Do not exceed 70 °C, or the solution may froth and overflow the beaker.
3. This demonstration is also an excellent method to show the relationship between temperature and reaction rate. Typically, an initial temperature of 50, 60, or 70 °C will produce a reaction time of 200, 90, or 40 s, respectively.
4. As a general rule, increasing the temperature of reaction by 10 degrees will double the rate of reaction.

QUESTIONS FOR STUDENTS

1. Write chemical equations for the reactions.
2. What is the formula of the activated complex?
3. What is the role of the hydrogen peroxide?
4. What gases are produced?
5. How can you be sure that the green color is due to the activated complex?

62.   Autocatalysis

A graduated cylinder contains a blue solution. A few drops of acid are placed on top of the solution, and in a few seconds a yellow layer appears. Within a few minutes, the yellow layer gradually moves down the column as the catalyzed reaction proceeds.

Procedure

Wear safety goggles.

1. Place 50 mL of water in a beaker.
2. Add 4 g of potassium chlorate, 12.5 g of sodium sulfite, and approximately 5 mg of bromophenol blue indicator. Do not mix these solids. Dissolve them in the water.
3. In a second beaker, add 4 mL of 3 M sulfuric acid to 50 mL of water.
4. Slowly, with constant stirring, add the diluted acid from the second beaker to the solution in the first beaker. Stir until everything dissolves. The solution should be blue–violet.
5. Fill a 100-mL graduated cylinder with the solution.
6. Carefully add two droppers full of 3 M sulfuric acid solution to the top of the liquid in the cylinder.
7. Soon a yellow color will appear at the top of the solution, and a yellow–blue interface will form.
8. Observe for several minutes as the yellow–blue interface moves down the graduated cylinder.

Reactions

1. This reaction is a redox reaction:

$$\underset{\text{blue}}{ClO_3^- \text{ (aq)}} + 3HSO_3^- \text{ (aq)} \rightarrow Cl^- \text{ (aq)} + \underset{\text{yellow}}{3SO_4^{2-} \text{ (aq)}} + 3H^+ \text{ (aq)}$$

2. The reaction proceeds only in an acidic solution.
3. Dropping sulfuric acid on the surface produces H^+ and other acidic products. These acidic products catalyze further reactions to produce additional acidic products. This is the autocatalytic effect.
4. Bromophenol blue indicator is yellow in highly acidic solutions. Thus, as autocatalysis proceeds, the blue color of the indicator changes to yellow.
5. The blue solution has a pH between 6.5 and 7.0 because of the buffering effect of the bisulfite–sulfite ions.

Solution

Sulfuric acid: Prepare a 3 M solution by adding 5 mL of concentrated sulfuric acid to 18 mL of distilled water. Use this acid only in steps 3 and 6. Caution: Use a face shield and disposable gloves when you handle concentrated H_2SO_4.

Teaching Tips

NOTES

1. The amount of bromophenol blue indicator is not critical. An amount about the size of a match head will work.
2. For best results, use anhydrous analytical-grade chemicals.
3. Point out that the acid merely begins the reaction. If the color change were due to this small amount of acid alone, the entire contents of the cylinder would immediately turn yellow.
4. Many biochemical reactions are autocatalytic. Pepsinogen is activated by hydrogen ions in the stomach to form pepsin. Pepsin then catalyzes the conversion of pepsinogen to additional pepsin.

QUESTIONS FOR STUDENTS

1. What is meant by autocatalysis?
2. Can you think of other autocatalytic reactions?
3. In this redox reaction, what is oxidized? What is reduced?
4. What happened at the interface between the yellow and blue layers?

63.   The Starch–Iodine Clock Reaction

Each of two beakers contains a clear solution. The solutions are mixed by pouring from one beaker into the other. After a few seconds, the mixed solution suddenly turns dark blue. Changing the concentration or the temperature of the solutions changes the time required for the blue color to be produced.

Procedure

Wear safety goggles.

1. Place 50 mL of solution A in a 250-mL beaker.
2. Place the same volume of solution B in a second beaker.
3. Mix the two solutions by pouring from one beaker into the other twice, then hold the filled beaker in full view of the class.
4. Note the time required for a reaction to occur after the solutions are mixed.
5. Repeat, but dilute solution A to half concentration. Note the time required for a reaction to occur.
6. Repeat, using solution A that has been warmed to 35 °C. Note the time required for a reaction to occur.

Reactions

The mechanism is not completely understood; however, the following simplified sequence is proposed:

1. IO_3^- reacts with HSO_3^- to form I^-:

$$IO_3^- + 3HSO_3^- \rightarrow I^- + 3H^+ + 3SO_4^{2-}$$

2. I^- reacts with IO_3^- to form I_2.
3. I_2 is immediately consumed by reaction with HSO_3^-:

$$I_2 + HSO_3^- + H_2O \rightarrow 2I^- + SO_4^{2-} + 3H^+$$

4. When all of the HSO_3^- has been used up, I_2 accumulates.
5. Iodine reacts with starch to form a colored complex:

$$I_2 + starch \rightarrow blue\ complex$$

Solutions

1. Solution A: 4.3 g of KIO_3 per liter.
2. Solution B: Make a paste of 4 g of soluble starch in a small amount of warm water. Slowly add 800 mL of boiling water. Boil for a few minutes, then cool the solution. Add 0.2 g of $Na_2S_2O_5$ (sodium metabisulfite). Add 5 mL of 1.0 M sulfuric acid (see Appendix 2). Dilute to 1 L.

Teaching Tips

NOTES

1. This demonstration allows you to illustrate beautifully the dependence of reaction rates on concentration and temperature.
2. Do not heat the solutions above 40 °C. The starch–iodine complex is unstable above 50 °C. Best results are obtained when the solutions are allowed to stabilize at room temperature for a few hours prior to mixing.
3. Ideally, 10–15 s should be required when the solutions are mixed at room temperature. If the reaction is too slow, add a little more sodium metabisulfite or more acid to solution B. If the reaction is too fast, dilute solution A.
4. $Na_2S_2O_5$ hydrolyzes in solution to $NaHSO_3$.
5. After sulfuric acid is added to solution B, it must be used within 10–12 h. If you need to keep the solution longer, add the acid just before using the solution.
6. The blue starch–iodine complex forms when the I_5^- or the I_3^- species fits inside the coiled amylose structure.

QUESTIONS FOR STUDENTS

1. Propose a simple mechanism for this reaction.
2. What is the role of the starch?
3. Why is there a delay before the reaction occurs?
4. What are the effects of concentration and temperature on reaction rates?
5. Why is this called a clock reaction?

64. The Old Nassau Clock Reaction

Three colorless solutions are mixed. In a few seconds, the solution turns bright orange, then suddenly turns dark blue.

Procedure

Wear safety goggles and disposable gloves. Mercury compounds are toxic.

1. Label three 250-mL beakers A, B, and C.
2. Pour 50 mL of solutions A, B, and C into their respective beakers.
3. Mix the solutions in this order: Add A to B to C.
4. Hold the beaker in view of the class.
5. Dispose of the solutions according to directions given in Appendix 5.

Reactions

$$IO_3^- + 3HSO_3^- \rightarrow I^- + 3SO_4^{2-} + 3H^+$$

$$Hg^{2+} + 2I^- \rightarrow HgI_2 \text{ (orange)}$$

$$6H^+ + IO_3^- + 5I^- \rightarrow 3I_2 + 3H_2O$$

$$I_2 + starch \rightarrow \text{blue complex}$$

Solutions

1. Solution A: Make a paste of 4 g of soluble starch in a few milliliters of warm water. Slowly add 500 mL of boiling water. Cool the solution, then add 15 g of $NaHSO_3$ and dilute to 1 L with distilled water.
2. Solution B: 3 g of $HgCl_2$ per liter distilled water. Wear gloves. Avoid contact with $HgCl_2$.
3. Solution C: 15 g of KIO_3 per liter distilled water.

Teaching Tips

NOTES

1. To speed up the reaction, use less of solution B.
2. The reaction is called the Old Nassau reaction because it produces the colors of Princeton University (orange and black). Nassau Hall is one of the older buildings on the Princeton campus. Hubert Alyea has delighted audiences throughout the world by performing this reaction as he sings the Princeton fight song.

QUESTIONS FOR STUDENTS

1. Propose a mechanism to explain how this reaction can produce two distinct colors.
2. How can the reaction rate be increased?
3. Is it necessary to mix the solutions in a particular order?
4. What compound is formed when the solution turns orange?

65.   Disappearing Orange Reaction: Now You See It, Now You Don't

Two beakers contain colorless solutions. A small amount of solution from beaker A is poured into beaker B and a bright orange color is produced. When the remainder of solution in beaker A is poured into beaker B, however, the orange color disappears.

Procedure

Wear safety goggles and disposable gloves.

1. Add equal volumes of solutions A and B to two large beakers labeled A and B.
2. Pour enough of solution A into solution B to produce a bright orange color. It will not take much of solution A.
3. Pour the remaining solution A into the B beaker and note that the orange color disappears.
4. Dispose of the solutions according to direction given in Appendix 5.

Reactions

$$Hg^{2+} + 2I^- \longrightarrow HgI_2 \text{ (orange)}$$

$$HgI_2 + 2I^- \longrightarrow HgI_4{}^{2-} \text{ (colorless)}$$

Solutions

1. Solution A: 5 g of potassium iodide in 300 mL of water.
2. Solution B: 2 g of mercuric chloride in 300 mL of water.

Teaching Tips

NOTES

1. KI forms an orange precipitate with $HgCl_2$. Excess KI dissolves this precipitate and forms a colorless complex.
2. For a variation of this reaction, see Demonstration 64, The Old Nassau Clock Reaction.

QUESTIONS FOR STUDENTS

1. Write the chemical equations for the reactions.
2. Is this a redox reaction?
3. What could you do to change the rate of the reaction?
4. How can you explain the disappearance of the orange color?

66.   A Traffic Light Reaction

A flask containing a pale yellow solution is gently swirled. The solution turns red. The flask is shaken, and the solution turns green.

Procedure

Wear safety goggles.

1. Place 50 mL of solution A in a 250-mL flask.
2. Add 5–10 mL of indicator solution. At the beginning of the demonstration, the solution should be light yellow.
3. Stopper the flask.
4. Gently swirl the flask to produce the red color.
5. Give the flask a quick shake to produce the green color.

Reactions

1. The indicator is reduced by alkaline dextrose, and a yellow color is produced.
2. When the flask is swirled, oxygen is added, the indicator is oxidized, and the red color is produced.
3. Shaking the flask introduces even more oxygen and causes further oxidation of the indicator to the green color.
4. Upon standing, the dextrose reduces the indicator back to the yellow color.

Solutions

1. Solution A: 3 g of dextrose (glucose) and 5 g of NaOH in 250 mL of water.
2. The indigo carmine indicator is a 1.0% solution.

Teaching Tips

NOTES

1. If the red color does not persist, increase the number of drops of indicator.
2. This traffic light has an advantage over others: It does not require a magnetic stirrer.
3. For a variation of this reaction, see Demonstration 76, The Blue Bottle Reaction.
4. Do not exceed the volume of Solution A recommended.

QUESTIONS FOR STUDENTS

1. Propose a chemical equation for the reaction.
2. Is this a redox reaction? If so, what is oxidized and what is reduced?
3. What role does the indicator play?
4. What happens when the flask is swirled? Shaken?
5. Will this reaction run down if the stopper remains in the flask?

67.   An Oscillating Reaction: Clear and Brown

A solution is stirred with a magnetic stirrer. The solution bubbles and fizzes, then turns brown, then colorless, and then brown again. The oscillations will continue for several minutes.

Procedure

Wear safety goggles; do this demonstration in a hood.

1. Place a large beaker on a magnetic stirrer.
2. Add 900 mL of water and 50 mL of concentrated sulfuric acid. Be careful: Use a face shield and disposable gloves when you handle concentrated H_2SO_4.
3. While the solution is stirring, add and dissolve the following: 3 spoons (about 20 g) of malonic acid, 2 spoons (about 20 g) of $KBrO_3$, and approximately ¼ spoon (about 3 g) of $MnSO_4$.
4. The solution will fizz and then turn brown.
5. Oscillations will begin after a few seconds.

Reactions

1. Oscillating reactions are complex. This one is thought to involve more than 20 chemical species and 18 steps in the reaction mechanism.
2. Products of this reaction include CO_2; formic acid, HCOOH; and bromomalonic acid, $BrCH(COOH)_2$.
3. Formation of different oxidation states of manganese results in the colorless and brown appearance of the solution as it oscillates.

Teaching Tips

NOTES

1. Because exact amounts of the chemicals used in step 3 are not important, we suggest that they be measured in plastic spoons. This will allow you to emphasize the dramatic oscillating effect.
2. An excellent article on oscillating reactions appears in *Scientific American* (March 1983, p 112).
3. If a magnetic stirrer is not available, gently stir the solutions in the beaker with a stirring rod. Avoid splashing.

QUESTIONS FOR STUDENTS

1. Why is stirring important in this reaction?
2. What causes the effervescence?
3. Why does the reaction eventually stop?
4. How could we renew the oscillations?

68. An Oscillating Reaction: Yellow and Blue

Solutions are mixed and placed on a magnetic stirrer. The color of the solution changes from light yellow to blue to light yellow. This oscillation will continue for 10–15 min.

Procedure

Wear safety goggles and disposable gloves.

1. Place 100 mL of solution A in a 500-mL beaker on a magnetic stirrer.
2. Set the stirrer on its slowest setting.
3. Add 100 mL of solution B.
4. Add 100 mL of solution C.
5. Oscillations will begin after a few seconds.

Reactions

1. This demonstration involves a series of complex reactions. In the first series of reactions, oxygen gas and iodine are formed.
2. The iodine reacts with starch to produce the blue color.
3. As the iodine is used up in another series of reactions, the color fades but is formed again when the concentration of iodine increases.

Solutions

1. Solution A: Add 40 mL of 30% hydrogen peroxide to 100 mL of water. Caution: Wear a face shield and gloves when you handle 30% H_2O_2.
2. Solution B: While stirring, add 4.3 g of KIO_3 and 0.5 mL of concentrated sulfuric acid to 100 mL of water.
3. Solution C: Prepare a paste of 0.15 g of soluble starch in hot water. While stirring, add this to 500 mL of hot water. Then add 7.8 g of malonic acid and 1.7 g of $MnSO_4 \cdot H_2O$.

Teaching Tips

NOTES

1. Notice that 30% hydrogen peroxide must be used. Be careful with this solution.
2. This reaction is similar to the starch–iodine clock in Demonstration 63.
3. Store 30% H_2O_2 in an approved refrigerator.

QUESTIONS FOR STUDENTS

1. What gas is produced in this reaction?
2. Propose a mechanism for this reaction.
3. Is this a redox reaction?
4. What might be done to renew the oscillation?

69.   An Oscillating Reaction: Red and Blue

Several solutions are placed in a Petri dish. After about 5 min, the color of the solution oscillates between red and blue.

Procedure

Wear safety goggles.

1. Prepare the oscillating solution as follows: Place 6 mL of solution A in a small beaker. Add 0.5 mL of solution B. Add 1.0 mL of solution C. A brown color will appear. When it disappears, add 1.0 mL of ferroin. Add 1 drop of Photoflo or some other surface-active (or wetting) agent.
2. Add enough of this solution to a Petri dish to fill it halfway.
3. Wait for oscillations to begin.

Reactions

1. Bromate reacts with malonic acid to produce bromomalonate.
2. Bromate also reacts with red ferrous dye to produce blue ferric dye.
3. Bromide and malonic acid react to form bromomalonate.
4. Bromomalonate and blue ferric dye react to form bromide.
5. Bromide inhibits the reaction of red ferrous dye to blue dye, and a red color is produced.

Solutions

1. Solution A: Dissolve 5 g of sodium bromate in 67 mL of distilled water. Then add 2 mL of concentrated sulfuric acid slowly, while stirring. Use gloves and a face shield when you handle concentrated H_2SO_4.
2. Solution B: 1 g of sodium bromide in 10 mL of distilled water.
3. Solution C: 1 g of malonic acid in 10 mL of distilled water.
4. Ferroin: 0.25 M solution.

Teaching Tips

NOTES

1. Ferroin is phenanthroline ferrous sulfate.
2. Photoflo can be found at any photography shop. It is a surface-active agent used in developing and printing.
3. This demonstration is adapted from a reaction described by Briggs and Raucher in *Scientific American* (July 1978).
4. This oscillating reaction does not require a magnetic stirrer.

QUESTIONS FOR STUDENTS

1. What is the role of ferroin in this reaction?
2. Can you propose a mechanism for this reaction?
3. Is this a redox reaction?
4. Are other ferrous salts red? Are ferric salts blue?

70.   Metathesis (Double Replacement) Reaction Between Two Solids

Two white solids are placed in a small beaker and shaken or stirred with a glass rod. A yellow solid is formed.

Procedure

Wear safety goggles and disposable gloves. Lead salts are toxic.

1. Place an approximately equal amount of lead nitrate and potassium iodide crystals in a small beaker or Petri dish.
2. Point out to students that both of these substances are white crystals.
3. Mix the two compounds in the beaker with a glass stirring rod, or, if they are in a Petri dish, cover the dish and gently swirl it.
4. Notice the appearance of a yellow product.
5. Discard the solid waste according to directions given in Appendix 5.

Reaction

$$Pb(NO_3)_2 \ (s) + 2KI \ (s) \rightarrow PbI_2 \ (s) + 2KNO_3 \ (s)$$
$$\text{white} \qquad\qquad \text{white} \qquad\quad \text{yellow}$$

Teaching Tips

NOTES

1. This demonstration offers a good opportunity to discuss the importance of surface area in chemical reactions.
2. You might want to pass the containers among the students and let them stir the mixture and see the PbI_2 that formed.
3. You can add equal volumes of KI and $Pb(NO_3)_2$ solutions to obtain the bright yellow PbI_2 precipitate.
4. You may use a well-stoppered large test tube to mix the two solids.

QUESTIONS FOR STUDENTS

1. How can a chemical reaction occur between two solids?
2. Write the equation for this reaction.
3. Show how this reaction represents a double replacement.
4. Is a yellow compound the only product of the reaction?

Oxidation-Reduction

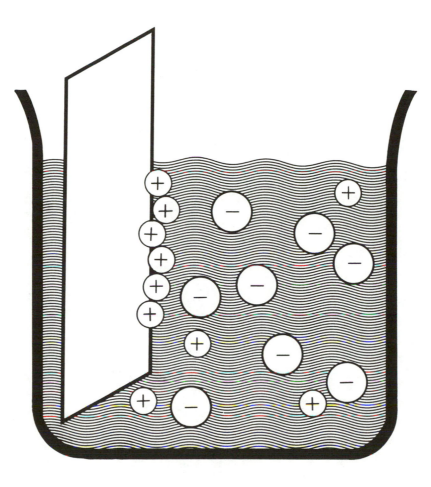

71.   Catalytic Oxidation of Ammonia

A copper coil is heated in the flame of a Bunsen burner until it is glowing hot. The coil is suspended above a layer of ammonium hydroxide in a flask. The coil continues to glow and may eventually get hot enough to melt.

Procedure

Wear safety goggles and do this demonstration in a hood.

1. Prepare a copper coil approximately 1 in. in diameter by winding about 1 ft of bare copper wire (24 gauge works well) around a test tube. Form a hook on the end of the wire.
2. Place about 50 mL of concentrated ammonium hydroxide in a 250-mL wide-mouth flask or beaker.
3. Heat the wire with a burner until it is glowing red-hot. Be careful.
4. Immediately place the wire in the flask, just above the level of the ammonium hydroxide.
5. Hook the end of the wire on the side of the flask.
6. Observe the reaction.
7. If the wire does not continue to glow, begin again, using ammonium hydroxide heated to 25 °C or add a small amount of sodium hydroxide solution.

Reaction

1. The hot copper causes ammonia to be oxidized with oxygen from the air:

$$4NH_3 \text{ (g)} + 3O_2 \text{ (g)} \xrightarrow{\text{Cu}} 2N_2 \text{ (g)} + 6H_2O \text{ (l)} + heat$$

2. The reaction is highly exothermic and may produce enough heat to melt the copper wire.

Teaching Tips

NOTES

1. Notice that bits of molten copper may fall into the solution. The blue color is due to the formation of copper complex ion.
2. Darken the room for a more spectacular effect.
3. You can also use a copper screen instead of a copper wire. It produces a more intense glow.
4. Wrap an old penny (pre-1982) with copper wire. Heat the wire and penny and suspend it over methyl alcohol. Be careful: Alcohol is flammable.
5. Ammonium hydroxide is more properly called aqueous ammonia.

QUESTIONS FOR STUDENTS

1. What is the catalyst in this reaction?
2. What gas is produced in the reaction?
3. How hot does the reaction get?
4. Is this a redox reaction? What is oxidized? What is reduced?

72.   Ammonia as a Reducing Agent

A mixture of black copper(II) oxide and ammonia is heated in a spectacular reaction to produce red copper(I) oxide and metallic copper.

Procedure

Wear safety goggles and do this demonstration in a hood.

1. Grind 1.0 g of ammonium carbonate with a mortar and pestle and mix it with 2.0 g of copper(II) oxide.
2. Place the mixture in a large test tube, spread the mixture so that it has the greatest surface area possible, position the tube horizontally, and clamp it into place on a ring stand.
3. Heat the mixture evenly with a hot burner flame until the reaction ceases.
4. Notice the production of metallic copper, red copper(I) oxide, and water at the mouth of the tube.

Reactions

Ammonia is a good reducing agent. It is made available as a reagent by the thermal decomposition of ammonium carbonate:

$$(NH_4)_2CO_3 \text{ (s)} \rightarrow 2NH_3 \text{ (g)} + CO_2 \text{ (g)} + H_2O \text{ (g)}$$

Copper(II) oxide is a good oxidizing agent. It reacts with ammonia to produce copper(I) oxide. Further reduction of the oxide produces metallic copper. Nitrogen gas is also a product of the oxidation of ammonia. A general reaction can be shown as

$$2Cu^{2+} + 2OH^- + 2e^- \text{ (reducing agent)} \xrightarrow{\text{heat}} Cu_2O \text{ (s)} + H_2O \text{ (g)}$$

Teaching Tips

NOTES

1. You can collect the nitrogen produced by the reaction. Place a stopper and delivery tube in the test tube and collect the gas by displacing water.
2. Ammonium carbonate decomposes readily with heat to produce ammonia. Ammonia is a good heart stimulant, so it has been used in smelling salts.
3. Copper(I) oxide (a reddish-brown solid) is produced when a reducing sugar, like glucose, reacts with a solution of copper(II) ions. This is called Benedict's test.
4. The oxidizing ability of copper(II) oxide has been used to determine the amount of carbon in certain organic compounds. The copper(II) oxide oxidizes carbon to carbon dioxide, which is then trapped in a previously weighed amount of sodium hydroxide. The increase in weight of sodium hydroxide gives a quantitative indication of the amount of carbon dioxide produced, hence the amount of carbon in the sample.
5. Try placing 10 g of ammonium carbonate on a top-loading balance. Observe the decrease in mass as the ammonium carbonate decomposes to ammonia and carbon dioxide, a spontaneous reaction at room temperature.

QUESTIONS FOR STUDENTS

1. What is the source of the ammonia in this reaction?
2. What properties of copper(II) oxide and copper(I) oxide allow you to tell them apart?
3. What has happened to the copper atoms in copper(II) oxide when copper(I) oxide is produced? Where do the electrons come from?

73.   Oxidation of Glycerin by Permanganate

A small amount of crystalline substance is placed in a pile, and a small amount of liquid is added. After a few seconds, a large puff of smoke and intense violet flames are produced.

Procedure

Wear safety goggles and do this demonstration in a hood. Have a fire extinguisher handy.

1. Place a small pile of granular potassium permanganate (about a tablespoon, 10–15 g) in the center of an evaporating dish.
2. Add 10 drops of glycerin on top of the pile.
3. Stand back: The reaction will occur in 15–20 s. This reaction produces a flame and intense heat.
4. Dissolve the solid waste in water and filter it. Flush the filtrate down the drain and place the precipitate in the solid waste container.

Reaction

$$14KMnO_4 \text{ (g)} + 4C_3H_5(OH)_3 \text{ (l)} \rightarrow 7K_2CO_3 \text{ (s)} + 7Mn_2O_3 \text{ (s)} + 5CO_2 \text{ (g)} + 16H_2O \text{ (l)}$$

Materials

1. Glycerin (glycerol) can be purchased at drugstores.
2. Use fresh, granular potassium permanganate ($KMnO_4$).

Teaching Tips

NOTES

1. Do not use more than the specified amount of $KMnO_4$.
2. Take special care with this reaction and all other reactions that involve rapid oxidation.
3. You can tell when the reaction is about to begin because a slight puff of smoke forms in the center of the pile.
4. Keep in mind that this reaction is a delayed reaction.
5. Add just enough glycerin to wet the top of the pile of $KMnO_4$.

QUESTIONS FOR STUDENTS

1. Is this a redox reaction? If so, what is the oxidizer?
2. Does the size of the permanganate particles influence the rate of the reaction?
3. Write the equation for this reaction.
4. What is the gas produced?

74. Oxidation–Reduction: Iron

Colorless solutions become blood red when oxidized. These red solutions again become colorless when reduced.

Procedure

Wear safety goggles.

Oxidation

1. Label two beakers 1 and 2.
2. Place 50 mL of solution A in each beaker.
3. Add 10 mL of solution B to each beaker.
4. Add 10 mL of the following oxidizing agents to show formation of the red Fe(III) complex: beaker 1, $KMnO_4$; and beaker 2, H_2O_2.

Reduction

Add $SnCl_2$ solution to either of the beakers until the solution becomes colorless [forms Fe(II) complex]. The reaction may take 30–60 s.

Reactions

Oxidation

$$Fe^{2+} \text{ (aq)} + \text{oxidizing agent} \longrightarrow Fe^{3+} \text{ (aq)}$$
$$\text{colorless}$$

$$Fe^{3+} \text{ (aq)} + SCN^- \text{ (aq)} \rightleftharpoons FeSCN^{2+} \text{ (aq)}$$
$$\text{red}$$

1. Beaker 1: Permanganate ion, MnO_4^-, is reduced to Mn^{2+} in acid solution.
2. Beaker 2: Peroxide is reduced to H_2O or OH^- in acid solution.

Reduction

$$Fe^{3+} \text{ (aq)} + Sn^{2+} \text{ (aq)} \longrightarrow Fe^{2+} \text{ (aq)} + Sn^{4+} \text{ (aq)}$$
$$\text{red} \qquad \text{reducing} \qquad \text{colorless}$$
$$\text{agent}$$

Solutions

1. Solution A: Dissolve 2 g of $(NH_4)_2SO_4 \cdot FeSO_4 \cdot 6H_2O$ (ferrous ammonium sulfate) in 20 mL of 6 M H_2SO_4 and dilute to 500 mL. Use a face shield and gloves when you handle 6 M H_2SO_4.
2. Solution B is 1 M potassium thiocyanate: Dissolve 97 g of KSCN per liter of solution.
3. The potassium permanganate solution is 0.05 M: Dissolve 7.5 g of $KMnO_4$ per liter of solution.
4. The tin(II) chloride solution is 0.1 M: Dissolve 19 g of $SnCl_2$ per liter of solution.
5. The hydrogen peroxide, H_2O_2, solution is the 3% variety available in drugstores.

Teaching Tips

NOTES

1. This demonstration shows the action of several oxidizing agents, as well as a reducing agent.
2. Cl_2 can also be used as an oxidizing agent. See Demonstration 7, Preparation of Chlorine Gas from Laundry Bleach, for preparation of Cl_2 gas.
3. These reactions project well. Use Petri dishes and an overhead projector.
4. These solutions work best if they are prepared just before use.
5. See Demonstrations 32 and 88 for variations of this reaction.

QUESTIONS FOR STUDENTS

1. Write complete equations for the reaction produced in each demonstration.
2. How does Sn^{2+} act as a reducing agent?
3. What else might act as an oxidizing agent for this reaction?
4. Can you detect any differences in the oxidizing power of the various oxidizing agents?

75. The Silver Mirror Reaction

Several solutions are placed in an Erlenmeyer flask. The flask is swirled and a silver mirror forms to coat the inside of the flask.

Procedure

Caution: The ammoniacal silver solution formed as a result of this demonstration is potentially explosive upon standing. Follow the procedure for properly disposing of this waste solution. Mix the solutions *only* in the order given in the procedure. Wear safety goggles and disposable gloves and use a safety shield.

1. Prepare to handle the waste product by adding 5 mL of 3 M HCl to a beaker; place a filter paper in a funnel for filtration.
2. Scrupulously clean a 50-mL Erlenmeyer flask.
3. Place 10 mL of solution A in the flask.
4. Mix 5 mL of solution B with 5 mL of solution C and add this mixture to the flask. Caution: Do not mix these solutions until the demonstration is performed.
5. Quickly add 10 mL of solution D to the flask.
6. Stopper the flask and mix it with a quick but gentle swirling motion. Cover the entire glass surface evenly with the solution while swirling. Continue swirling until the silver mirror forms.
7. Immediately pour the waste silvering solution from the flask into the beaker containing acid prepared in step 1.
8. Immediately rinse the silvered flask several times with water, flushing the washings down the sink.
9. Filter the precipitate in the beaker. Flush the filtrate down the drain and dispose of the solid silver chloride precipitate according to directions given in Appendix 5.

Reaction

Metallic silver is formed when silver ion oxidizes the aldehyde part of a sugar molecule (glucose).

(aldehyde)

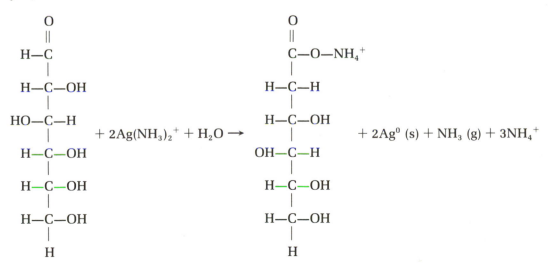

Solutions

1. Solution A: Dissolve 2.5 g of glucose (dextrose) and 2.5 g of fructose in 50 mL of water. Add 0.6 g of tartaric acid. Bring the solution to a boil, then cool it. Add 10 mL of ethyl alcohol and dilute to 100 mL.
2. Solution B: Dissolve 0.4 g of silver nitrate in 50 mL of distilled water.
3. Solution C: Dissolve 6.0 g of ammonium nitrate in 50 mL of water.
4. Solution D: Dissolve 10 g of sodium hydroxide in 100 mL of water. Use a safety shield when you handle NaOH solution.

Teaching Tips

NOTES

1. You can use a large, round-bottom flask, but you must swirl it quickly to cover the glass surface. Rinse it thoroughly.
2. This demonstration shows the reduction of silver ion to silver metal and the reducing ability of sugars containing free aldehyde groups.
3. If the mirror does not form, the flask was probably not clean.
4. The ammoniacal silver solution produced by this reaction on standing is potentially explosive. It must be decomposed and disposed of immediately.
5. If desired, remove the mirror by dissolving it in concentrated HNO_3. Be careful.
6. Adding the waste silvering solution to HCl precipitates waste silver as $AgCl_2$. This can be disposed of as a heavy metal waste.
7. The hydrogen on the aldehyde group was removed from the sugar molecule when it was oxidized.

QUESTIONS FOR STUDENTS

1. Write a chemical equation for the reaction.
2. What is the silver mirror?
3. Do you think that this method is a good way to produce a mirror? Is it economically feasible?
4. What is the role of the sugar in this reaction?
5. Which hydrogen was removed from the sugar molecule when it was oxidized?

76.   The Blue Bottle Reaction

A large round-bottom flask contains a colorless solution. When shaken vigorously, a blue color forms. After a few seconds, the blue color fades and the solution again becomes colorless. The process can be repeated.

Procedure

Wear safety goggles.

1. Show your students a stoppered flask containing approximately 300 mL of solution.
2. Give the flask a few vigorous shakes. Notice that the solution turns blue.
3. After a few seconds, the solution becomes clear again.
4. Ask for suggestions to explain these reactions.
5. The process can be repeated several times. It may be necessary to remove the stopper periodically.

Reactions

1. The students should deduce that the reaction causing the blue color is essentially

$$\text{gas} + \text{liquid} \longrightarrow \text{blue color}$$

and that the clearing of the solution is simply

$$\text{blue color} + x \longrightarrow \text{colorless}$$

2. The actual reaction involves the reduction of methylene blue by an alkaline dextrose solution. Upon shaking, the reduced product is reoxidized to the blue dye. Essentially, the reaction occurs in four steps:

$$O_2 \text{ (g)} + O_2 \text{ (dissolved)}$$

$$\underset{\text{colorless}}{\text{methylene blue}} \xrightarrow{\text{dissolved } O_2} \underset{\text{blue}}{\text{methylene blue}}$$

$$\text{glucose} + OH^- \longrightarrow \text{glucoside}$$

$$\text{glucoside} + \underset{\text{blue}}{\text{methylene blue}} \longrightarrow \underset{\text{colorless}}{\text{methylene blue}} + OH^-$$

Solution

Prepare the solution for the flask as follows: Add 8 g of KOH to 300 mL of water. Cool the solution and add 10 g of glucose (dextrose). Add a few drops of methylene blue indicator solution, or a small amount of the solid indicator (no larger than a match head). Do not add too much.

Teaching Tips

NOTES

1. This classic demonstration is a favorite of our colleague, J. A. Campbell, and is discussed in his book, *Why Do Chemical Reactions Occur?* (Prentice–Hall: Englewood Cliffs, NJ, 1965).
2. This reaction is an excellent inquiry demonstration. Challenge your students to suggest why the blue color is produced and why it later disappears.
3. Students should be expected to deduce the overall reaction, but not the formation of the glucoside.
4. The solution is good for only about 12–15 shakings.
5. To show the students that nothing on the stopper causes the reaction, remove the stopper and shake the flask; the blue color still appears.
6. The interface between the gas and the liquid is where the reaction occurs.
7. If the debluing were simply a reverse of the bluing, gas would be evolved. A reverse reaction does not occur unless conditions are changed. In this demonstration, conditions remain the same.
8. Shaking the solution dissolves oxygen.

QUESTIONS FOR STUDENTS

1. How do you know that a reaction occurred?
2. Could something on the stopper cause the reaction?
3. What do you notice at the gas–liquid interface?
4. Could this interface be where the reaction is occurring?
5. Is the debluing simply a reverse of the bluing?
6. What does the shaking do?
7. Why does the reaction eventually run down?

77.   Oxidation of Zinc: Fire and Smoke

A drop of water is added to a small pile of chemicals in an evaporating dish. After a few seconds, an intense blue flame with smoke is produced.

Procedure

Wear safety goggles. Do this demonstration behind a safety shield in a hood. Have a fire extinguisher handy.

1. Mix (do not grind) ammonium nitrate and ammonium chloride in a 4:1 ratio to give enough mixture to half fill a small evaporating dish.
2. Sprinkle enough zinc dust on the mixture to cover its surface lightly.
3. Add about 3 drops of water from a dropper, letting the water run down the side of the dish.
4. Stand back: A flame and intense heat will be produced. There may be spattering; keep students at least 5 ft away.
5. Dissolve the solid waste in water and filter it. Flush the filtrate down the drain and place the precipitate in the solid waste container.

Reactions

1. Cl^- (from NH_4Cl) acts as a catalyst on the decomposition of NH_4NO_3:

$$NH_4NO_3 \text{ (s)} \xrightarrow{\text{Cl}^-} N_2O \text{ (g)} + 2H_2O \text{ (aq)}$$

2. Water produced in the reaction causes the decomposition of more NH_4NO_3, which is an autocatalytic effect.
3. The reaction melts the NH_4NO_3 and allows the oxidation of the zinc. The overall reaction is probably as follows:

$$Zn \text{ (s)} + NH_4NO_3 \text{ (s)} \rightarrow N_2 \text{ (g)} + ZnO \text{ (s)} + 2H_2O \text{ (g)}$$

Teaching Tips

NOTES

1. The chemicals in this reaction represent a mixture of oxidizing and reducing agents. Water acts as a catalyst.
2. This reaction is highly exothermic.
3. This reaction produces a dense cloud of white ZnO (s). You can add a few crystals of iodine to produce purple smoke. Perform this reaction in a hood.
4. The oxidizing properties of nitrates are demonstrated.

QUESTIONS FOR STUDENTS

1. Write the chemical equation for the reaction.
2. What is the catalyst? What is autocatalysis?
3. What was oxidized? What was reduced?
4. What is the smoke?
5. What was the cause of the short delay before this reaction began?

78. An Improved Mercury Beating Heart

A small amount of mercury is placed in a solution in a watch glass. When touched with an iron wire, the mercury pulsates in a rhythmic fashion, resembling a beating heart.

Procedure

Wear safety goggles and disposable gloves. Perform this demonstration in a hood. Avoid breathing mercury vapor.

1. Place a large watch glass inside a Petri dish.
2. Add clean mercury to the watch glass to form a pool no more than ¾ in. in diameter.
3. Add dilute sulfuric acid until the surface of the mercury pool is just covered.
4. Place a clean, small iron wire on the watch glass so that it barely touches the mercury. Fix the wire in position with a bit of putty.
5. Carefully drop hydrogen peroxide on top of the mercury.
6. The mercury will soon begin to pulsate.
7. To get a stronger beat, slowly add hydrogen peroxide, drop by drop, and adjust the iron wire for maximum effect.
8. After the demonstration, rinse the mercury several times, then store it until you need it again.

Reactions

1. The Hg pool forms a sphere because of a large electrical charge, that is, number of electrons, on the surface.
2. H_2O_2 acts as an oxidizing agent. Electrons are removed from the Hg, and the Hg drop flattens.
3. When the drop flattens, it touches the wire and receives electrons.

$$H_2O_2 \text{ (aq)} + 2H^+ + 2e^- \longrightarrow 2H_2O \text{ (l)}$$

$$Fe \text{ (aq)} \longrightarrow Fe^{3+} \text{ (aq)} + 3e^-$$

4. The increased number of electrons on the Hg causes it to become spherical again, and it moves away from the wire.
5. The process is repeated.

Solutions

1. The dilute sulfuric acid is less than 1.0 M: 1 mL of concentrated H_2SO_4 in 17 mL of water.
2. The H_2O_2 is 6%; either dilute 5 mL of 30% H_2O_2 in 15 mL of water or use Clair-oxide, a 6% H_2O_2 solution available in drugstores. Use a face shield and gloves when you handle concentrated H_2O_2. Store it in an approved refrigerator.

Teaching Tips

NOTES

1. Reagent-grade Hg works best. It must be pure and clean.

2. The watch glass is placed in a Petri dish so that it can be moved easily and to catch any accidental spillage.

3. Put the Petri dish on the overhead projector.

4. You will see several patterns of oscillations (beats): a small to large sphere, an equilateral triangle (the most common type), or a four-lobed shape.

5. A very thin iron wire works well.

6. Be prepared to clean up any spills. See Appendix 5 for the proper procedure.

7. This method has several advantages over the standard dichromate method: The beat is stronger and easier to begin, there is no colored chromate residue, and it does not require the use of carcinogenic chromates.

QUESTIONS FOR STUDENTS

1. Explain what is happening when the mercury pool contracts and expands.

2. Why is mercury an ideal substance to use in this demonstration?

3. Is a catalyst used in this reaction?

4. Does the mercury settle into a regular beating pattern? Why?

79.   Oxidation States of Manganese: Quick Mn^{6+}

A solution of purple Mn^{7+} is poured over a folded white paper napkin and is immediately reduced to the green Mn^{6+}.

Procedure

Wear safety goggles.

1. Place 20 mL of potassium permanganate solution in a small beaker.
2. Add a few milliliters (the amount is not critical) of sodium hydroxide solution.
3. Mix the two solutions. Note the purple color that is characteristic of the Mn^{7+} ion.
4. Fold a heavy white paper napkin several times and pour the solution over the napkin.
5. Note the immediate and intense green color of the Mn^{6+} ion on the napkin. After a minute or so, brown Mn^{4+} may also begin to form on the napkin.

Reaction

The cellulose in the napkin reduces the alkaline Mn^{7+} to Mn^{6+}. Further reduction produces Mn^{4+}.

Solutions

1. The potassium permanganate is 0.01 M: Dissolve 1.6 g of $KMnO_4$ per liter.
2. The sodium hydroxide solution is 1 M: Dissolve 40 g of NaOH per liter.

Teaching Tips

NOTES

1. See Demonstration 80 for ways to show other oxidation states of manganese.
2. Get white napkins from the school cafeteria. Filter paper will work well also.
3. This reaction is similar to that shown in Demonstration 76. The cellulose in paper is a polymer of dextrose molecules.

QUESTIONS FOR STUDENTS

1. Will this reaction occur if the solution is poured over a white handkerchief instead of a napkin?
2. What does the napkin do?
3. Write an equation to show this reaction.
4. Does the green color persist?

80. Oxidation States of Manganese: Mn^{7+}, Mn^{6+}, Mn^{4+}, and Mn^{2+}

Three reactions are carried out with potassium permanganate, producing colors characteristic of the various oxidation states of manganese.

Procedure

Wear safety goggles and disposable gloves. Use a face shield when using H_2SO_4 and NaOH.

1. Label four large beakers A, B, C, and D.
2. Place 50 mL of $KMnO_4$ solution in each beaker.
3. Set beaker D aside; it represents manganese in the $+7$ oxidation state.
4. Add 30 mL of H_2SO_4 to beaker A.
5. Add 40 mL of NaOH to beaker C.
6. Place beaker A on a white background. Slowly add $NaHSO_3$ solution while stirring. Note the change of color to red, pink, and finally colorless. This colorless solution indicates the $+2$ oxidation state.
7. Place beaker B on the same background and add $NaHSO_3$ while stirring. Note the formation of a brown precipitate, indicating the $+4$ state.
8. Place beaker C on the same background and add $NaHSO_3$ while stirring. Note the formation of a green color, indicating the $+6$ state.
9. Arrange the four beakers to show the colors and the $+7$, $+6$, $+4$, and $+2$ oxidation states.

Reactions

1. In beaker A ($+7$ to $+2$)

$$2MnO_4^- \text{ (aq)} + H^+ \text{ (aq)} + 5HSO_3^- \text{ (aq)} \rightleftharpoons 2Mn^{2+} \text{ (aq)} + 5SO_4^{2-} \text{ (aq)} + 3H_2O \text{ (l)}$$
purple   colorless

2. In beaker B ($+7$ to $+4$)

$$OH^- + 2MnO_4^- \text{ (aq)} + 3HSO_3^- \text{ (aq)} \rightleftharpoons 2MnO_2 \text{ (s)} + 3SO_4^{2-} \text{ (aq)} + 2H_2O \text{ (l)}$$
purple   brown

3. In beaker C ($+7$ to $+6$)

$$2MnO_4^- \text{ (aq)} + 3OH^- \text{ (aq)} + HSO_3^- \text{ (aq)} \rightleftharpoons 2MnO_4^{2-} \text{ (aq)} + SO_4^{2-} \text{ (aq)} + 2H_2O \text{ (l)}$$
purple   green

Solutions

1. The potassium permanganate solution is 0.01 M: Dissolve 0.74 g of $KMnO_4$ per 500 mL.
2. The H_2SO_4 solution is 1 M (see Appendix 2).
3. The sodium hydroxide solution is 1.0 M: dissolve 4 g of NaOH per 100 mL.
4. The sodium bisulfite solution is 0.01 M: Dissolve 0.54 g of $NaHSO_3$ per 500 mL.

Teaching Tips

NOTES

1. Use sodium bisulf*ite*, not sodium bisulf*ate*.
2. MnO_4^- (+7) is purple; MnO_4^{2-} (+6) is green; MnO_2 (+4) is a brown precipitate; and Mn^{2+} is colorless to light pink.
3. Manganese has oxidation states not demonstrated here, such as violet +3 ion in Mn_2O_3 and $MnCl_3$.

QUESTIONS FOR STUDENTS

1. Write the chemical equation for these reactions.
2. List the different oxidation states of manganese and their colors.
3. Does manganese have oxidation states not demonstrated here?
4. Are some oxidation states easier to obtain than others?
5. Indicate what was oxidized and what was reduced in each reaction.

81.   The Prussian Blue Reaction

Two colorless solutions are mixed. After a few seconds the mixture turns yellow, then green, and finally a deep blue (Prussian blue).

Procedure

Wear safety goggles.

1. Mix equal volumes of solutions A and B in a large beaker.
2. Observe the color changes.

Reactions

1. The first reaction produces the yellow hexacyanoferrate(II) ion, $Fe(CN)_6^{4-}$.
2. The second reaction involves the mixing of the yellow ion with the blue $Fe(III)[Fe(II)(CN)_6]$ species to produce a green solution.
3. The final reaction is the production of the blue complex, $KFe[Fe(CN)_6]$.
4. The overall reaction is

$$K^+ \text{ (aq)} + [Fe(CN)_6]^{4-} \text{ (aq)} + Fe^{3+} \text{ (aq)} \longrightarrow KFe[Fe(CN)_6]$$
$$\text{yellow} \qquad\qquad\qquad\qquad\qquad\qquad \text{Prussian blue}$$

Solutions

1. Solution A is 0.0002 M $K_4Fe(CN)_6 \cdot 3H_2O$. Obviously, this solution is quite dilute. Prepare it by dissolving 0.422 g of potassium ferrocyanide in 100 mL of water to make a 0.01 M solution. Dilute this solution to 0.0002 M by diluting 1 mL of solution to 50 mL.
2. Solution B is 0.0002 M $NH_4Fe(SO_4)_2 \cdot 12H_2O$. Note: This solution is not in water; it is a solution in potassium bisulfate. First prepare a 0.01 M solution by dissolving 0.48 g of ammonium ferric sulfate in 100 mL of 0.1 M $KHSO_4$ solution. Dilute 1 mL of this solution in 50 mL of the $KHSO_4$ solution to prepare the 0.0002 M solution.
3. Potassium hydrogen sulfate solution is 0.1 M. Dissolve 2.72 g of $KHSO_4$ in 200 mL of water.

Teaching Tips

NOTES

1. Proper concentration is essential for forming all three colors. Keep trying until you get distinct and sharp color changes.
2. This method is a colorful way to show a reaction in a series of steps before the final oxidation–reduction product is obtained.
3. This reaction is responsible for the production of blueprints. See Demonstration 86 for this application of the Prussian blue reaction.
4. Prussian blue has the formula $KFe[Fe(CN)_6]$.

QUESTIONS FOR STUDENTS

1. Write the chemical equations for the reactions.
2. Which of the observed reactions is indicative of the rate of the overall reaction?
3. Are some reaction rates different? Why?
4. Determine the oxidizing and reducing agents.

82.   Copper into Gold: The Alchemist's Dream

A copper penny is placed in an evaporating dish and heated with a mixture. It turns silver. The penny is then heated on a hot plate, and it suddenly turns gold.

Procedure

Wear safety goggles, gloves, and a face shield. Note step 10: *special disposal procedure.*

1. Place approximately 5 g of zinc dust in an evaporating dish.
2. Add enough NaOH solution to cover the zinc and fill the dish about one-third.
3. Place the dish on a hot plate and heat until the solution is near boiling.
4. Prepare a copper penny (pre-1980) by cleaning it thoroughly with a light abrasive (steel wool pads work well).
5. Using crucible tongs or tweezers, place the cleaned penny in the mixture in the dish.
6. Leave the penny in the dish for 3–4 min. You will be able to tell when the silver coating is complete.
7. Remove the penny, rinse it, and blot dry with paper towels. (Do not rub.) Remove all particles of zinc.
8. Using crucible tongs or tweezers, place the coated penny on the hot plate. The gold color appears immediately.
9. When the gold color forms, remove the coin, rinse it, and dry it with paper towels.
10. *Special disposal procedure:* Do not discard the waste zinc in the trash container. When zinc dries, it forms a powder that may spontaneously ignite. Rinse the NaOH–zinc mixture several times with water. Then add the solid to a beaker that contains 200 mL of 1 M H_2SO_4. When all of the solid dissolves, flush the zinc sulfate solution down the drain.

Reactions

1. The first reaction is the plating of the copper with zinc: Zinc reacts with sodium hydroxide to form a sodium zincate, $[Zn(OH)_3(H_2O)]^-Na^+$. This reaction gives the silver color to the penny.
2. The second reaction is the formation of the brass alloy. This alloy gives the penny the gold color. Heat causes a fusion of the zinc and copper.

Solution

The sodium hydroxide solution is 6 M: Dissolve 240 g of NaOH per liter. Wear safety goggles, gloves, and a face shield.

Teaching Tips

NOTES

1. You may not want to deny your students the opportunity to perform this demonstration as an experiment; they love it. They must wear safety goggles, gloves, and a face shield.

2. New pennies seem to work best if they are not overheated. Pre-1980 pennies are copper; later pennies have a zinc core.

3. This demonstration can be used to illustrate a variety of reactions: solid–solid reactions, oxidation–reduction reactions, or metallurgical reactions.

4. Brass is 60–82% Cu and 18–40% Zn.

5. With care, you can use a burner instead of a hot plate. Do not overheat the penny.

QUESTIONS FOR STUDENTS

1. Is this reaction an oxidation–reduction reaction?
2. Why did the penny turn "silver"?
3. Why did it turn "gold"?
4. Why did we heat the penny to turn it "gold"?

83.   Oxidation of Sodium

A small piece of sodium metal is dropped into a water-filled Petri dish on an overhead projector. The sodium forms a small ball and darts around the dish, leaving a pink trail.

Procedure

Wear safety goggles.

1. Place the smaller half of a Petri dish on an overhead projector.
2. Half fill the dish with water.
3. Add 1 drop of phenolphthalein and 1 grain of laundry detergent.
4. Stir the contents of the dish thoroughly.
5. Using tweezers, place a small piece of freshly cut sodium metal (not larger than 8 mm^3) in the dish.
6. Cover the dish with the other half of the Petri dish.
7. Project and observe the random motion of the sodium and the color change in the solution.

Reactions

1. The sodium is oxidized:

$$Na \text{ metal (s)} \rightarrow Na^+ \text{ (aq)} + e^-$$

2. The water is reduced:

$$2H_2O \text{ (l)} + 2e^- \rightarrow H_2 \text{ (g)} + 2OH^- \text{ (aq)}$$

3. The basic solution reacts with the phenolphthalein to produce the pink trail behind the sodium.

Teaching Tips

NOTES

1. This reaction is an excellent way to demonstrate acid–base reactions, oxidation–reduction reactions, the action of an indicator, and the activity of metallic sodium.
2. Do not use larger pieces of sodium. An explosion can occur when large pieces of sodium are placed in water.
3. The dish is covered to prevent spattering on the projector glass.
4. If the sodium gets stuck on the side of the dish, give it a push with the tweezers to keep it moving.
5. Do not substitute any other group I metal except lithium. Do *not* use potassium, rubidium, or cesium.

QUESTIONS FOR STUDENTS

1. Why was detergent added?
2. How does the sodium behave?
3. How can you explain this behavior on a chemical basis?
4. What causes the pink trail?

84.   Oxidation States of Vanadium: Reduction of V^{5+} to V^{2+}

A flask containing a yellow solution is gently shaken. The solution turns blue. The flask is shaken again and the solution turns green. More vigorous shaking produces a violet color.

Procedure

Wear safety goggles and disposable plastic gloves. Vanadium is toxic.

1. Place 50 g of zinc amalgam (see Solutions) in a 1-L Florence flask.
2. Add 200 mL of vanadium solution.
3. Stopper the flask.
4. Note the color of the solution. Pour a little of the solution in a beaker as a sample.
5. Gently swirl the flask and note the blue color. Save a small amount as a sample.
6. Shake the flask and note the green color. Save a sample.
7. Shake the flask more vigorously and note the violet color. Save a sample.
8. Dispose of the vanadium according to directions given in Appendix 5.

Reactions

Reaction 1

$$Zn\ (s) + 2VO_3^-\ (aq) + 8H_3O^+\ (aq) \rightleftharpoons 2VO^{2+}\ (aq) + Zn^{2+}\ (aq) + 12H_2O\ (l)$$

Reaction 2

$$Zn\ (s) + 2VO^{2+}\ (aq) + 4H_3O^+\ (aq) \rightleftharpoons 2V^{3+}\ (aq) + Zn^{2+}\ (aq) + 6H_2O\ (l)$$

Reaction 3

$$Zn\ (s) + 2V^{3+}\ (aq) \rightleftharpoons 2V^{2+}\ (aq) + Zn^{2+}\ (aq)$$

Summary

$$V^{5+}\ (aq) \rightarrow V^{4+}\ (aq)\ \text{(yellow to green)}$$

$$V^{4+}\ (aq) \rightarrow V^{3+}\ (aq)\ \text{(green to blue)}$$

$$V^{3+}\ (aq) \rightarrow V^{2+}\ (aq)\ \text{(blue to violet)}$$

Solutions

1. Zinc amalgam: In a 500-mL flask, dissolve 2 g of mercuric chloride in 300 mL of water. Add 2 mL of concentrated nitric acid. Add 250 g of zinc (20–30 mesh works best). Stopper the flask and shake it for a few minutes. Pour off the liquid and wash the amalgam several times with water. Store the amalgam under water in a sealed jar.
2. Vanadium solution: In a large beaker, dissolve 6–8 g of NaOH in 200 mL of water. Add 10 g of ammonium vanadate (NH_4VO_3). You may need to warm the beaker to dissolve this compound completely. Stir constantly. Add 500 mL of

2 M sulfuric acid solution. Dilute to 1 L. Store in a large, well-stoppered bottle. Wear goggles, gloves, and a face shield when you prepare the H_2SO_4 and NaOH solutions.

Teaching Tips

NOTES

1. These colors project well and can be used in Petri dishes on an overhead projector.
2. The used amalgam can be washed several times and reused.
3. Vanadium is named for Vanadis, the Norse goddess of beauty.

QUESTIONS FOR STUDENTS

1. What is an amalgam?
2. Write the equation for the reaction producing each colored vanadium ion.
3. What is oxidized and what is reduced in each reaction?
4. What does shaking do?

85. Oxidation States of Vanadium: Reoxidation of V^{2+} to V^{5+}

Demonstration 84 reduced vanadium through the +5, +4, +3, and +2 oxidation states. This demonstration reverses the process by reoxidizing the vanadium solution through the +2, +3, +4, and, finally, +5 states.

Procedure

Wear safety goggles and disposable gloves. Vanadium is toxic.

1. Fill a buret with ceric sulfate solution.
2. Add this solution drop by drop to the violet solution from Demonstration 84, swirling constantly.
3. Note the appearance of colors characteristic of the various oxidation states as more ceric sulfate solution is added.
4. Dispose of the vanadium according to the directions given in Appendix 5.

Reactions

Reaction 1

$$Ce^{4+} (aq) + V^{2+} (aq) \rightleftharpoons V^{3+} (aq) + Ce^{3+} (aq)$$

Reaction 2

$$Ce^{4+} (aq) + V^{3+} (aq) + 3H_2O (l) \rightleftharpoons VO^{2+} (aq) + Ce^{3+} (aq) + 2H_3O^+ (aq)$$

Reaction 3

$$Ce^{4+} (aq) + VO^{2+} (aq) + 6H_2O (l) \rightleftharpoons VO_3^- (aq) + Ce^{3+} (aq) + 4H_3O^+ (aq)$$

Summary

$$V^{2+} (aq) \longrightarrow V^{3+} (aq) \text{ (violet to blue)}$$

$$V^{3+} (aq) \longrightarrow V^{4+} (aq) \text{ (blue to green)}$$

$$V^{4+} (aq) \longrightarrow V^{5+} (aq) \text{ (green to yellow)}$$

Solutions

1. Vanadium solution: Use the solution prepared for Demonstration 84.
2. The ceric sulfate solution is 0.1 M: Add 33.2 g of $Ce(SO_4)_2$ to 100 mL of 1.0 M H_2SO_4 (see Appendix 2).

Teaching Tips

NOTES

1. Sulfuric acid is added to the ceric sulfate solution to increase the solubility of the salt.

2. You may want to do Demonstrations 84 and 85 together, but a lot of chemistry is involved.

QUESTIONS FOR STUDENTS

1. Write the reactions for the reoxidation of vanadium.
2. What color is characteristic of each oxidation state?
3. To what group of elements does vanadium belong?

86.   Photoreduction: The Blueprint Reaction

A key, or some similar object, is placed on a sheet of blue-green paper. After exposing this paper to a floodlight for a few minutes, the paper is washed and an imprint of the key is left on the paper.

Procedure

Wear safety goggles and disposable gloves.

1. Presoak a light-sensitive paper by coating ordinary bond paper with a photosensitive solution.
2. Allow the treated paper to dry in a dimly lit room.
3. Place an object such as a key, a comb, or a pencil on the paper.
4. Expose the paper to a strong light (a floodlight or sunlight) for 2–3 min.
5. Wash the exposed paper in running water.
6. Note the imprint.

Reactions

1. The photosensitive compound, iron(III)hexacyanoferrate(III), $Fe[Fe(CN)_6]$, is formed when the two solutions are mixed.
2. $Fe[Fe(CN)_6]$ is reduced by light to form iron(III)hexacyanoferrate(II), $Fe_4[Fe(CN)_6]_3$.
3. The overall reaction is
   a. Coating the paper:

$$Fe^{3+} (aq) + Fe(CN)_6^{3-} \rightleftharpoons \underset{\text{bronze-green}}{Fe[Fe(CN)_6] (aq)}$$

   b. Photoreduction on the paper:

$$Fe^{3+} \xrightarrow[\text{citrate}]{\text{light}} Fe^{2+}$$

$$3Fe^{2+} + Fe(CN)_6^{3-} \longrightarrow \underset{\text{blue}}{Fe_3[Fe(CN)_6]_2}$$

Solutions

1. Make two solutions:
   Solution A: Dissolve 30 g of potassium hexacyanoferrate(III), $K_3Fe(CN)_6$, in 100 mL of water.
   Solution B: Dissolve 40 g of iron(III) ammonium citrate in 100 mL of water.
2. Mix these two solutions in a dimly lit room (no direct sunlight) to form the photosensitive solution.

Teaching Tips

NOTES

1. You can easily coat the paper by using a sponge soaked in solution. Use a paper that does not soak up the solution.

2. For best results, keep the dry, treated paper in a drawer or box prior to its use.

3. Wash your hands immediately after treating the paper.

4. Place a black-and-white negative on the paper, cover both with a glass plate, and expose to strong light for about 5 min. When the paper is washed, a crude photograph will be produced.

5. Silver halides are reduced by light when you take a picture.

QUESTIONS FOR STUDENTS

1. Write the equation for the blueprint reaction.

2. What role does light play in this reaction?

3. What is the oxidation state of iron in each of the complexes formed in this reaction?

4. Can you give other examples of photoreduction reactions?

87.   Hydrogen Peroxide as an Oxidizing Agent: Black to White

A black precipitate of lead sulfide is formed. When a solution of hydrogen peroxide is added, a vigorous reaction results that produces oxygen gas and a white precipitate of lead sulfate.

Procedure

Wear safety goggles and disposable plastic gloves. Do this demonstration in a hood.

1. Place 10 mL of lead nitrate solution in a large test tube. Carefully add a few drops of ammonium sulfide solution to the test tube. Notice the immediate formation of a black precipitate. Caution: Use only a few drops of ammonium sulfide. Do not leave the bottle uncapped. Avoid contact with this solution.
2. Filter the solution, collecting the black precipitate in the filter paper. Wash several times with water. Discard the filtrate by pouring it down the sink and flushing with water.
3. Scoop some of the black precipitate into another large test tube.
4. Add 10 mL of hydrogen peroxide solution to the test tube.
5. Observe the bubbling of the mixture and the eventual formation of a white precipitate.
6. Dispose of the solid lead waste according to directions given in Appendix 5.

Reactions

Lead sulfide is formed by the lead nitrate and ammonium sulfide:

$$Pb^{2+} (aq) + S^{2-} (aq) \longrightarrow \underset{\text{black}}{PbS (s)}$$

Hydrogen peroxide oxidizes the sulfide to sulfate. The hydrogen peroxide is reduced to water.

$$\underset{\text{black}}{PbS (s)} + 4H_2O_2 (aq) \longrightarrow \underset{\text{white}}{PbSO_4} + 4H_2O$$

Solutions

1. Regular hydrogen peroxide (3%) from the drugstore works well.
2. Ammonium sulfide, $(NH_4)_2S$, is available as a liquid.
3. Lead nitrate solution is 0.1 M. Dissolve 16.5 g of lead nitrate, $Pb(NO_3)_2$, in 500 mL of water. Wear gloves when you prepare this solution.

Teaching Tips

NOTES

1. When an aqueous solution of ammonia is saturated with hydrogen sulfide, ammonium hydrogen sulfide (NH_4HS) is formed. If this solution is then treated with an equivalent amount of aqueous ammonia, ammonium sulfide, $(NH_4)_2S$, is formed.

2. In 1818, the French chemist Louis Jacques Thenard first prepared hydrogen peroxide by reacting hydrochloric acid and barium peroxide. Thenard worked with Gay-Lussac; together they discovered the element boron in 1808.

3. Ultraviolet light in the atmosphere reacts with oxygen mixed with water vapor to produce hydrogen peroxide, which appears in small amounts in rain and snow.

4. Hydrogen peroxide is used to treat sewage for control of hydrogen sulfide odors. It has also been used to restore the white color to old paintings where the original white lead pigment has been converted over many years to the black sulfide.

5. Hydrogen peroxide is a strong oxidizing agent, but it can also act as a reducing agent. It will reduce manganese dioxide and potassium permanganate with the rapid evolution of oxygen gas.

6. Generally, when hydrogen peroxide acts as an oxidizing agent, water is the only product; when it acts as a reducing agent, water and oxygen are products.

7. Hydrogen peroxide is used in different concentrations as an antiseptic, a bleach for hair, and an oxidizing agent for high explosives.

QUESTIONS FOR STUDENTS

1. What is the appearance of the lead sulfide and the lead sulfate?
2. Give the reactions for the oxidation of lead sulfate by hydrogen peroxide.
3. The hydrogen peroxide used in this experiment is 3%. Hydrogen peroxide also exists commonly in 6%, 9%, 30%, and even 90% solutions. What are some uses for these concentrations?

88.   Color Changes in Fe(II) and Fe(III) Solutions

Two sets of four small beakers are arranged on the demonstration table. A solution is poured from a bottle or jug into each beaker and an assortment of colored solutions is produced.

Procedure

Wear safety goggles and disposable plastic gloves. $BaCl_2$ is toxic.

Ferrous ion, Fe(II)

1. The beakers are prepared by adding a few crystals (approximately 0.5 g) of each of the following solids to approximately 10 mL of water in the beakers: potassium hexacyanoferrate(III) (potassium ferricyanide), $K_3Fe(CN)_6$; tannic acid, $C_{76}H_{52}O_{46}$; barium chloride, $BaCl_2$; and sodium hydrogen sulfite, $NaHSO_3$.
2. The bottle contains iron(II) ammonium sulfate solution, $Fe(NH_4)_2(SO_4)_2 \cdot 6H_2O$.
3. Pour enough solution from the bottle into each beaker to produce the color indicated: beaker 1, blue; beaker 2, black; beaker 3, white; and beaker 4, yellow.

Ferric ion, Fe(III)

1. The beakers contain a small amount of the following crystals and approximately 10 mL of water: potassium thiocyanate, KSCN; potassium hexacyanoferrate(II) (potassium ferrocyanide), $K_4Fe(CN)_6$; tannic acid, $C_{76}H_{52}O_{46}$; and sodium hydrogen sulfite, $NaHSO_3$.
2. The bottle contains iron(III) ammonium sulfate solution, $Fe(NH_4)(SO_4)_2 \cdot 12H_2O$.
3. Pour enough solution from the bottle into each beaker to produce the color indicated: beaker 1, red; beaker 2, blue; beaker 3, black; and beaker 4, yellow–orange.

Reactions

Ferrous ion, Fe(II)

$$3Fe^{2+} (aq) + 2Fe(CN)_6^{3-} (aq) \longrightarrow \underset{\text{blue}}{Fe_3[Fe(CN)_6]_2}$$

$$Fe^{2+} (aq) + \text{tannic acid} \longrightarrow Fe(II) \text{ tannate (aq)} \overset{[O]}{\longrightarrow} \underset{\text{black}}{Fe(III) \text{ tannate (aq)}}$$

$$SO_4^{2-} (aq) + Ba^{2+} (aq) \longrightarrow \underset{\text{white}}{BaSO_4 (s)}$$

Ferric ion, Fe(III)

$$Fe^{3+} (aq) + SCN^- (aq) \longrightarrow \underset{\text{red}}{Fe(SCN)^{2+}}$$

$$Fe^{3+} (aq) + [Fe(CN)_6]^{4-} (aq) \longrightarrow Fe^{2+} (aq) + \underset{\text{blue}}{[Fe(CN)_6]^{3-} (aq)}$$

$$Fe^{3+} (aq) + \text{tannic acid} \longrightarrow \underset{\text{black}}{Fe(III) \text{ tannate (aq)}}$$

$$Fe^{3+} (aq) + SO_4^{2-} (aq) \longrightarrow \underset{\text{yellow–orange}}{Fe_2(SO_4)_3}$$

Solutions

1. Iron(II) ammonium sulfate, $Fe(NH_4)_2(SO_4)_2 \cdot 6H_2O$ and iron(III) ammonium sulfate, $Fe(NH_4)(SO_4)_2 \cdot 12H_2O$ are dilute solutions. The concentration is not critical; 5–10 g per liter works well.
2. Try various amounts of solids to get best results.

Teaching Tips

NOTES

1. Tannic acid has the formula $C_{76}H_{52}O_{46}$.
2. Iron(II) ammonium sulfate is readily oxidized to the +3 state.
3. This demonstration is excellent for a chemical show. Make up a story to go with the color changes.
4. KSCN gives a red color with Fe(III) but not with Fe(II).

QUESTIONS FOR STUDENTS

1. Give the formula for the compounds responsible for each color.
2. Write the equations for each reaction.
3. How could you differentiate between the Fe(II) ion and the Fe(III) ion?
4. What can you learn about solubility from this demonstration?

89.   The Activity Series for Some Metals

A Petri dish is placed on an overhead projector. HCl is added to the dish and five metal pieces are placed in labeled areas. After a few seconds, bubbles of hydrogen gas appear on the metals. The hydrogen bubbles form at a rate relative to the activity of each metal.

Procedure

Wear safety goggles.

1. Outline the circumference of a glass Petri dish with a felt-tip marker on a plastic sheet. Label five areas in the circle on the sheet Cu, Fe, Sn, Zn, and Mg.
2. Place the plastic sheet on an overhead projector and place the dish over the drawn circle. The labeled symbols should project as if within the dish.
3. Cover the bottom of the dish with 6 M HCl.
4. Carefully place small pieces of Cu, Fe, Sn, Zn, and Mg near their respective labels.
5. Note the extent of reaction of each metal with HCl.

Reaction

The general reaction for a metal with an acid is

$$\text{metal (s)} + 2HCl \text{ (aq)} \longrightarrow H_2 \text{ (g)} + \text{metal Cl}^- \text{ (aq)}$$

Solutions

The HCl solution is 6 M (see Appendix 2). Wear disposable gloves and a face shield when you use 6 M HCl.

Teaching Tips

NOTES

1. This demonstration is an excellent way to view five reactions simultaneously.
2. The order, according to rate of hydrogen gas formation, is Mg > Zn > Fe > Sn > Cu.

QUESTIONS FOR STUDENTS

1. Write the chemical equation for each reaction.
2. What observed reaction is indicative of the rate of the reaction?
3. Which metal reacts most with HCl? Which metal reacts least?
4. Would a different acid change the activity order?
5. Would a different acid concentration change the activity order?

90.   Metal Trees

Various metals are placed in flasks containing clear solutions. After a few minutes metallic crystals form on the metal surface, often resembling branches on a tree.

Procedure

Wear safety goggles and disposable gloves; toxic metals are used. Here are several ways to demonstrate these reactions:

1. Silver tree: Place a heavy, coiled copper wire in a 2% silver nitrate solution. You can observe a black coating on the wire immediately. After an hour or so, beautiful silver crystals form. Avoid contact with the silver nitrate solution.
2. Tin tree: Place a strip of iron or a coil of iron wire in a container of tin chloride solution. Observe the formation of tin crystals.
3. Lead tree: Place a strip of zinc in a 5% solution of lead nitrate. Lead crystals form on the zinc strip.
4. Dispose of the metal salts according to the directions given in Appendix 5.

Reactions

$$Ag^{2+} \text{ (aq)} + Cu \text{ (s)} \longrightarrow Ag \text{ (s)} + Cu^{2+} \text{ (aq)}$$

$$Sn^{2+} \text{ (g)} + Fe \text{ (s)} \longrightarrow Sn \text{ (s)} + Fe^{2+} \text{ (aq)}$$

$$Pb^{2+} \text{ (aq)} + Zn \text{ (s)} \longrightarrow Pb \text{ (s)} + Zn^{2+} \text{ (aq)}$$

Solutions

1. The silver nitrate solution is 2%: Dissolve 4 g of $AgNO_3$ in 200 mL of distilled water.
2. The tin chloride solution is 5%: Dissolve 10 g of $SnCl_2$ in 200 mL of water.
3. The lead nitrate solution is 5%: Dissolve 10 g of $Pb(NO_3)_2$ in 200 mL of water.

Teaching Tips

NOTES

1. Each of these reactions involves the formation of a metal by a displacement reaction.
2. Oxidation-reduction should also be pointed out to the students.
3. Although the silver tree is beautiful and classic, silver is quite expensive. Use it sparingly and take necessary precautions when using silver nitrate.
4. Although a reaction is apparent in a few minutes in each of these reactions, complete crystal formation may take several hours.

QUESTIONS FOR STUDENTS

1. Write the equation for each reaction demonstrated.
2. Can you predict what will happen in each reaction?
3. How are these reactions related to the electromotive series of elements?
4. Devise a demonstration to show tree formation from a metal and a solution.

91.   The Aluminum Cola Can Rip-Off

With a quick, twisting motion, an aluminum cola can is ripped into two pieces.

Procedure

Wear safety goggles.

1. Prepare several aluminum cola cans for the demonstration as follows:
   a. Insert a triangular file through the opening of an empty aluminum cola can. With a smooth, circular motion, carefully scratch through the thin plastic coating inside the can. Continue until you have scratched completely around the circumference of the can.
   b. Fill the can with copper chloride solution.
   c. After 3–5 min, pour out the solution and rinse the can.
2. If the cans were prepared properly, only the paint on the outside of the can will be holding the two halves together. Hold the top and bottom of the can in each hand and pull the halves apart with a quick, twisting motion.

Reactions

In this oxidation–reduction, copper ion is reduced to metallic copper and aluminum metal is oxidized to the aluminum ion.

$$3Cu^{2+} \text{ (aq)} + 2Al \text{ (s)} \longrightarrow 3Cu \text{ (s)} + 2Al^{3+} \text{ (aq)}$$

Solution

Copper chloride solution is 1.0 M. Dissolve 100 g of $CuCl_2$ per liter of solution.

Teaching Tips

NOTES

1. Notice the deposits of copper metal on the ends when two halves of the can are pulled apart. If any copper chloride is spilled on top of the can, copper will be deposited there also.
2. A fun variation of the demonstration is to mix several treated cans with several untreated cans and invite students to try to pull them apart.
3. Aluminum beverage cans came into use early in the 1970s. Now, more than 2 million tons of aluminum is used annually in aluminum cans.
4. Aluminum is well suited for use in beverage cans: The metal is odorless, nontoxic, lightweight, and tasteless. It is also a good thermal conductor, and aluminum beverage cans can be cooled quickly.
5. The thin plastic coating on the inside of the aluminum can and the paint on the outside prevent it from forming aluminum oxide.
6. The durability of the aluminum can has created an environmental pollution problem. More than 4 million tons of aluminum cans, foil, and containers are discarded each year. Fortunately, much of this is recycled.

QUESTIONS FOR STUDENTS

1. What is oxidized in this demonstration? What is reduced?
2. What properties of aluminum make it suitable for food containers?
3. How is aluminum prepared commercially?

92.   Displacement of Tin by Zinc

Mossy zinc is added to a solution in a beaker. After a few seconds, spongy tin is formed and floats to the surface of the liquid.

Procedure

Wear safety goggles, disposable gloves, and a face shield. Do this demonstration in a hood.

1. Place 200 mL of $SnCl_2$ solution in a 600-mL beaker.
2. Add 40 mL of concentrated HCl and stir the solution.
3. Sprinkle about a dozen pieces of granular mossy zinc into the solution; cover the bottom of the beaker with zinc.
4. Spongy tin immediately begins to form and then rises to the surface.

Reactions

This reaction nicely demonstrates two simultaneous reactions:

1. Tin is formed by displacement:

$$Zn\ (s) + SnCl_2\ (aq) \longrightarrow Sn\ (s) + ZnCl_2\ (aq)$$

2. It rises to the surface because hydrogen is also formed:

$$Zn\ (s) + 2HCl\ (aq) \longrightarrow H_2\ (g) + ZnCl_2\ (aq)$$

Solutions

1. The $SnCl_2$ solution is 10%: Dissolve 20 g of $SnCl_2$ in 200 mL of water.
2. The HCl is concentrated. Use gloves and a face shield when you handle concentrated HCl.

Teaching Tips

NOTES

1. The $SnCl_2$ may not be a clear solution, but it will be clear when HCl is added.
2. For a slower reaction, use dilute HCl.
3. Have samples of Zn and Sn available for students to examine and compare with the product.
4. If you want to repeat the demonstration, remove the spongy tin with tongs. (Be careful: It may have soaked up concentrated HCl.) Add a few crystals of $SnCl_2$ and a few milliliters of concentrated HCl.

QUESTIONS FOR STUDENTS

1. Write the chemical equations for these reactions.
2. What is the electrochemical reason for the reaction?
3. The reverse reaction cannot occur spontaneously. Why not?
4. Why did the tin float to the surface?

93.   Making a Simple Battery: The Gerber Cell

A simple battery is made from a large baby food jar. When completed, the battery will generate 1.5 V and light a small bulb. Six of these batteries connected in series will operate a 9-V pocket radio.

Procedure

Wear safety goggles.

1. Construct a battery according to Figure 4. Use a large baby food jar and a one-hole rubber stopper (size #9). A beaker works as well.

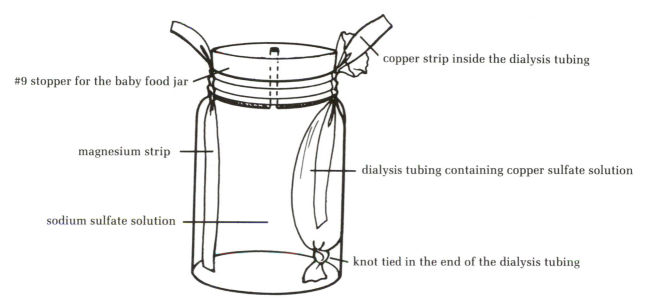

copper strip inside the dialysis tubing

#9 stopper for the baby food jar

magnesium strip

dialysis tubing containing copper sulfate solution

sodium sulfate solution

knot tied in the end of the dialysis tubing

Figure 4. A simple battery made from a baby food jar.

2. Cut a strip of copper metal long enough to fit into the jar; allow approximately 1 in. extra. Cut a magnesium strip the same length.
3. If the copper strip is not shiny, dip it into a dilute HNO_3 solution for a few seconds. Rinse it thoroughly in running water.
4. Cut a 6-in. length of dialysis tubing and hold it under water until it becomes flexible. Tie a knot in one end to make a bag. Insert the copper strip in the dialysis bag and fill the bag with copper sulfate solution. Place the prepared bag in the jar.
5. Clean the magnesium strip by dipping it quickly into 1 M HCl. Place the magnesium strip in the jar and fill the jar with sodium sulfate solution.
6. Insert the stopper so that the metal strips and dialysis bag are held in place.
7. Observe the reaction.
8. Attach wire leads to the metal strips and connect these to a flashlight bulb.
9. Prepare five more cells and connect the six cells in series. Use wire leads with alligator clips. They will operate a 9-V toy, radio, or calculator for several hours.

Reaction

1. This battery operates because of a transfer of electrons between magnesium and copper.

$$Mg \text{ (s)} + Cu^{2+} \text{ (aq)} \rightarrow Mg^{2+} \text{ (aq)} + Cu \text{ (s)} + 1.5 \text{ V}$$

2. Electrolysis of water produces bubbles of hydrogen and oxygen gas. Therefore, the stopper should have a hole in it.
3. A green coating of copper will begin to appear on the magnesium strip after the cell has operated for a while.

Solutions

1. The copper sulfate solution is 0.5 M: Dissolve 80 g of copper sulfate in 1 L of water.
2. The sodium sulfate solution is 0.5 M: Dissolve 71 g of sodium sulfate in 1 L of water.

Teaching Tips

NOTES

1. Dialysis tubing is a common supply in biology laboratories.
2. Wet the dialysis tubing before tying off the end and filling the sack with solution.
3. Mg is oxidized at the anode (+); Cu is reduced at the cathode (–).
4. You can use small beakers. The stopper serves to keep the metal strips in place.

QUESTIONS FOR STUDENTS

1. Trace the flow of electrons in this reaction.
2. Is this an oxidation–reduction reaction? If so, what is oxidized and what is reduced?
3. What reactions are evident in the cell?
4. Which metal strip is the anode and which is the cathode?

94.   Synthesis of Mercury(I) Iodide and Mercury(II) Iodide

A stoichiometric amount of iodine is ground with a mortar and pestle with metallic mercury, producing the green powder, mercury(I) iodide. More iodine is added and mixed, producing the red–yellow mercury(II) iodide.

Procedure

Wear safety goggles and disposable gloves; mercury is toxic.

1. Weigh two 6.35-g samples of iodine that has been ground to a fine powder. Caution: Do not inhale iodine.
2. Weigh a 10.0-g sample of metallic mercury and place it in a large mortar. Be careful: Be prepared to clean up any mercury spills.
3. Carefully grind the two together with a pestle until a green powder is produced.
4. Add the other iodine sample and continue grinding the mixture until a red–yellow powder is produced.
5. Dispose of the solid waste according to the directions given in Appendix 5.

Reactions

1. Metallic mercury reacts with iodine to produce mercury(I) iodide:

$$2Hg \text{ (s)} + I_2 \text{ (s)} \longrightarrow \underset{\text{green}}{Hg_2I_2 \text{ (s)}}$$

   Although we often write the formula of this compound as HgI, it is actually Hg_2I_2 because the mercury(I) ion is actually a dimercury ion, with two mercury atoms joined by a covalent bond, Hg_2^{2+}.
2. Mercury(I) iodide reacts with additional iodine to produce mercury(II) iodide:

$$Hg_2I_2 \text{ (s)} + I_2 \text{ (s)} \longrightarrow \underset{\text{red-yellow}}{2HgI_2 \text{ (s)}}$$

Teaching Tips

NOTES

1. Be careful when you handle metallic mercury. Wear plastic gloves for the demonstration. If any mercury is spilled, scoop it up and place it in a stoppered glass container for indefinite storage. (See Appendix 5.)
2. Go through the equations with students to show them that 6.35-g samples of iodine represent the stoichiometric amounts needed to react completely with 10 g of mercury.
3. You can also show that iodide ions will precipitate mercury(II) ions from solution as red HgI_2. This red compound will dissolve in an excess of iodide ions to form the complex tetraiodomercurate(II) ions, a colorless solution.

$$HgI_2 \text{ (s)} + 2I^- \longrightarrow (HgI_4)^{2-} \text{ (aq)}$$

QUESTIONS FOR STUDENTS

1. Show the reaction for the production of mercury(I) iodide.
2. What evidence is there that chemical changes have occurred in these two reactions?
3. Why is mercury(I) iodide written as Hg_2I_2?

95. Separating Iodine from Iodized Salt

Potassium iodide is removed from a sample of salt by dissolving the KI in alcohol. It is then oxidized with a hydrogen peroxide solution to iodine and dissolved in petroleum ether to produce a faint pink color.

Procedure

Wear safety goggles. Do this demonstration in a hood. Extinguish all flames when you use ethyl alcohol.

1. Place about 20 g of iodized salt in a 250-mL Erlenmeyer flask. Add 25 mL of ethyl alcohol, stopper it, and shake the flask vigorously to dissolve the KI. Let the flask sit for 5–10 min, occasionally shaking it.
2. Filter the solution into an evaporating dish and evaporate the filtrate to dryness. Alcohol is flammable. This evaporation must be done with a steam bath or with a hot plate in a hood.
3. Add 5 mL of 3% hydrogen peroxide (H_2O_2) to the evaporating dish and warm it slightly to dissolve the residue.
4. Carefully transfer the solution to a test tube and add 1–2 mL of a nonpolar solvent such as petroleum ether.
5. Stopper and shake the test tube and note the pink color of the petroleum ether solution, indicating the presence of a small amount of iodine.
6. Prepare a standard test tube by adding 5 mL of petroleum ether to 10 mL of hydrogen peroxide solution. Compare any color produced in this test tube with that in the petroleum ether layer of the other test tube. Add a few drops of a potassium or sodium iodide solution and observe the formation of iodine and the pink color in the petroleum ether layer.
7. Dispose of the petroleum ether according to the directions given in Appendix 5.

Reactions

The iodide ion, which is present in iodized salt as potassium iodide, dissolves in ethyl alcohol, whereas the salt does not readily dissolve. H_2O_2 oxidizes iodide to molecular iodine. Molecular iodine dissolves in the organic solvent to give the pink color.

Solutions

1. Hydrogen peroxide, H_2O_2, is 3%. The variety from the drugstore works well.
2. NaI or KI solution can be of any concentration.

Teaching Tips

NOTES

1. Use iodized table salt directly from the box.
2. Iodine was discovered when it was extracted from seaweed in 1811. Seaweed ash contains about 1% iodide ion.
3. Iodized table salt contains 0.01% KI or NaI.

4. The thyroid gland needs about 1 mg of iodine per week to synthesize the growth hormone, thyroxine. A deficiency of this hormone causes an enlargement of the thyroid gland, called a goiter.

5. The solubility of NaI in ethyl alcohol is 42.6 g per 100 mL of alcohol, whereas NaCl is only slightly soluble in alcohol.

6. Petroleum ether is a mixture of lightweight hydrocarbons.

7. A table of reduction potentials shows that the oxidation potential for I^- is -0.53 V. Other potentials indicate that H_2O_2 in acid could oxidize $2I^-$ to I_2 (1.77 V), or bleach, Cl_2 (aq) (1.36 V), or Fe^{3+} (aq) (0.77 V).

QUESTIONS FOR STUDENTS

1. What property allows I^- to be separated from Cl^- in table salt?

2. Look up the solubilities of the various ion combinations (salts) in any handbook of chemistry and physics.

3. What oxidizing agents might be used to oxidize $2I^-$ to I_2?

Colloids

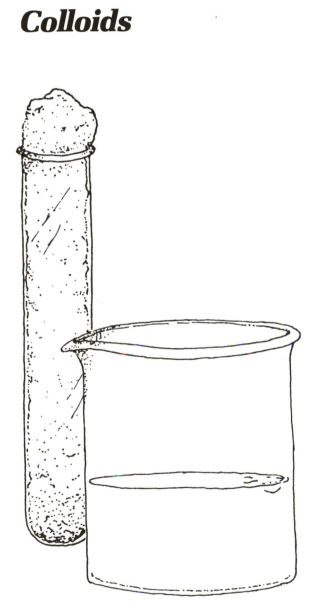

96. A Chemical Sunset

A Petri dish is fitted into a cardboard mask on an overhead projector. A solution is added and the lights are dimmed. The white light projected through the solution gradually turns yellow, then red, and finally opaque, a change simulating the colors seen during the sunset.

Procedure

Wear safety goggles.

1. Cut a hole the size of a Petri dish in a piece of cardboard large enough to cover the top of an overhead projector.
2. Place the Petri dish in the hole.
3. Add enough $Na_2S_2O_3$ solution to cover the bottom of the dish.
4. Add about 5 mL of concentrated HCl and quickly stir the solution. Wear gloves and a face shield when you use concentrated HCl.
5. Observe the color changes.

Reactions

The reaction produces colloidal sulfur that scatters light as it is being formed and produces different colors. A natural sunset is observed when light is scattered by dust particles in the atmosphere.

1. HCl reacts with sodium thiosulfate to produce thiosulfuric acid.

$$2H^+ \text{ (aq)} + S_2O_3^{2-} \text{ (aq)} \rightarrow H_2S_2O_3 \text{ (aq)}$$

2. Thiosulfuric acid decomposes immediately, producing sulfurous acid and sulfur in a colloidal suspension.

$$H_2S_2O_3 \rightarrow H_2SO_3 + \text{colloidal sulfur}$$

3. As the amount of colloidal sulfur increases, more light is blocked and the various colors are produced.

Solutions

1. The sodium thiosulfate pentahydrate solution, $Na_2S_2O_3 \cdot 5H_2O$ is 0.03 M: Dissolve 7 g per liter of water.
2. The hydrochloric acid is concentrated.

Teaching Tips

NOTES

1. Sodium thiosulfate pentahydrate is also known as hypo and is used in photography.
2. The formation of colloidal sulfur should take about 25–30 s.
3. You can also perform this demonstration by placing the solutions in a glass container in the beam of light from a slide projector.

4. Try decomposing other thiosulfates and polysulfides with acids to produce the
   sunset effect.

QUESTIONS FOR STUDENTS

1. What is a colloidal suspension?
2. How is this reaction similar to a natural sunset?
3. What causes the different colors to be produced?
4. Write a chemical equation for this reaction.
5. What other compounds might produce a similar reaction?

97. Production of Sterno: A Gel

Two beakers contain clear solutions. When the contents of one is poured into the other, a solid gel immediately forms.

Procedure

Wear safety goggles. Caution: Ethyl alcohol is flammable; extinguish all flames.

1. Place 5 mL of solution A in a small beaker.
2. Place 30 mL of solution B in a second beaker.
3. Challenge a student to see how many times the contents of the two beakers can be mixed; pour A into B.
4. When the solutions are first mixed, they immediately form a gel. This gel is Sterno.

Reactions

The structure of this gel is unclear. The calcium acetate probably forms a network that traps the ethyl alcohol molecules.

Solutions

1. Solution A: The calcium acetate is a saturated solution. Dissolve 35 g of calcium acetate, $Ca(CH_3COO)_2$, in 100 mL of warm water.
2. Solution B: 100% ethyl alcohol works best, but other alcohols also work.

Teaching Tips

NOTES

1. If the gel does not form immediately, you need more calcium acetate in solution.
2. A gel is a colloidal system consisting of a liquid (alcohol) dispersed in a solid (calcium acetate).
3. Other gels include jelly, gelatin, and agar (the bacterial growth medium used in biology laboratories).
4. To show that this product really is canned heat, light the beaker. Caution: The flame is almost colorless, but the heat is intense. Turn off the lights to see the pale blue flame better.
5. Try varying the ratio of calcium acetate and alcohol for maximum effect.

QUESTIONS FOR STUDENTS

1. What is a gel?
2. How is a gel, such as canned heat, produced?
3. What are some properties of this gel?

98.   Production of a Foam

Two clear solutions are mixed and a chemical foam is produced.

Procedure

Wear safety goggles.

1. Place 50 mL of solution A in a 250-mL beaker.
2. Place 50 mL of solution B in a second beaker.
3. Pour the contents of A into B and mix quickly.
4. Invert the beaker to show the stability of the foam.

Reactions

1. This foam is produced by the action of carbon dioxide gas on a detergent solution.
2. Aluminum sulfate, $Al_2(SO_4)_3$ provides the acid component for the demonstration:

$$[Al(H_2O)_x]^{3+} \text{ (aq)} + H_2O \text{ (l)} \rightleftharpoons H_3O^+ \text{ (aq)} + [Al(OH)(H_2O)_{x-1}]^{2+} \text{ (aq)}$$

3. $NaHCO_3$ produces HCO_3^-:

$$NaHCO_3 \text{ (s)} \longrightarrow Na^+ \text{ (aq)} + HCO_3^- \text{ (aq)}$$

4. The products of reactions 2 and 3 react to produce CO_2 gas:

$$HCO_3^- \text{ (aq)} + H_3O^+ \text{ (aq)} \longrightarrow 2H_2O \text{ (l)} + CO_2 \text{ (g)}$$

Solutions

1. Solution A: Place 1.0 g of laundry detergent and 7.0 g of $Al_2(SO_4)_3 \cdot 18H_2O$ in a mortar and grind it into a powder. Dissolve the powder in 50 mL of water.
2. Solution B: Dissolve 5.0 g of $NaHCO_3$ in 50 mL of water.

Teaching Tips

NOTES

1. A chemical foam contains CO_2; a mechanical foam contains air.
2. A foam is a colloidal system with a gas dispersed in a liquid.
3. Other foams include whipped cream and shaving cream.

QUESTIONS FOR STUDENTS

1. What reactions led to the production of the foam?
2. How is this reaction similar to that involving the production of CO_2 during the baking process?
3. Describe the foam.
4. Name some other examples of foams.

99.   Another Foam

Two clear solutions are mixed and a foam is produced.

Procedure

Wear safety goggles.

1. Place 100 mL of aluminum sulfate solution in a large beaker or a large graduated cylinder.
2. Add 100 mL of albumin–sodium bicarbonate solution.

Reactions

1. CO_2 is produced by the action of the acidic aluminum sulfate on the sodium bicarbonate.
2. The CO_2 is trapped by the egg white, and a foam is formed.

Solutions

1. Aluminum sulfate solution: Dissolve 25 g of $Al_2(SO_4)_3$ in 100 mL of water.
2. Albumin–sodium bicarbonate solution: Add 25 g of sodium bicarbonate and 2 g of egg albumin in 150 mL of warm water. Add the albumin slowly. Stir and heat until most of the solids dissolve. Cool, decant the top 100 mL of solution, and discard the remainder.

Teaching Tips

NOTES

1. A foam is a dispersion of a gas in a liquid.
2. Soap suds can also be used to trap a foam; albumin makes a more stable foam than soap suds.
3. See Demonstrations 60 and 98 for other methods of producing foams.

QUESTIONS FOR STUDENTS

1. What is a foam?
2. How was a foam produced in this reaction?
3. What is the role of the albumin?
4. Name some other foams.

Polymers

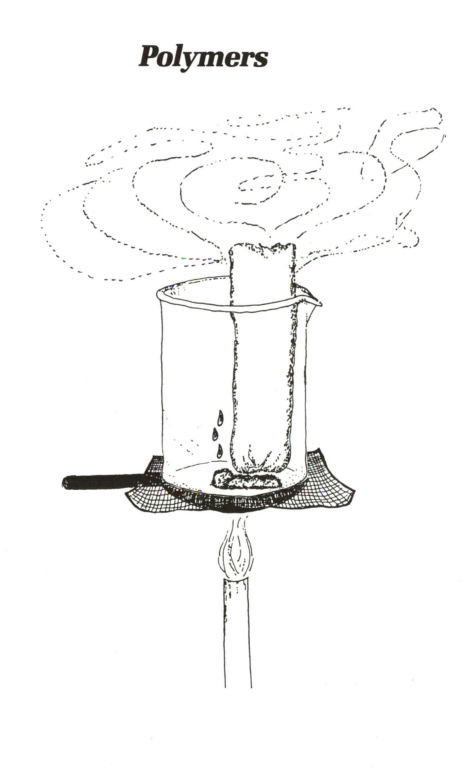

100.   Dehydration of *p*-Nitroaniline: Snake and Puff

A small amount of sulfuric acid is added to a small amount of solid in a beaker. After heating the beaker a few seconds, a giant, sausagelike snake pops out of the beaker with a great deal of smoke. This demonstration is spectacular. However, it produces a copious amount of smoke that the class may find irritating. It should be performed only outdoors or in a hood.

Procedure

Wear safety goggles and disposable gloves. This demonstration must be done outdoors or in a hood. Wear a face shield when you work with concentrated H_2SO_4.

1. Prepare a paste using a small amount of concentrated sulfuric acid and *p*-nitroaniline.
2. Place a small beaker containing the paste on a wire gauze and heat gently with a burner.
3. When the material begins to char and bubble, reduce the heat and stand back.

Reaction

The sulfuric acid dehydrates the *p*-nitroaniline and produces sulfur dioxide and long, plasticlike polymer snakes.

Teaching Tips

NOTES

1. Do not substitute any other compound for *p*-nitroaniline.
2. Avoid breathing *p*-nitroaniline. Its vapors are toxic.

QUESTIONS FOR STUDENTS

1. What is the appearance of the residue? What does it look like?
2. What is the probable element in the residue?
3. What does dehydration mean?
4. What is the gas that helped form the snake?
5. Why do you suppose the snake was produced suddenly, rather than a little at a time?

101.   Synthesis of Nylon

Two liquids are mixed in a small beaker and nylon is formed at their interface. The nylon is pulled from the beaker; a continuous thread 10–15 ft long is formed.

Procedure

Wear safety goggles and disposable gloves. Do this demonstration in a hood.

1. Place 5 mL of solution A in a small beaker.
2. Place 5 mL of solution B in a second beaker.
3. Slowly add solution A to solution B by pouring it down the side of the beaker. Do not stir or mix.
4. A film will form at the interface of the two solutions.
5. Carefully hook the film with a bent paper clip and pull the film from the beaker.
6. Continue pulling until the solutions are exhausted.

Reactions

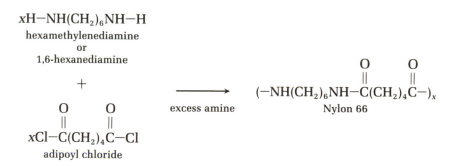

Solutions

1. Solution A: Prepare a 0.5 M basic solution of hexamethylenediamine (or 1,6-diaminohexane) as follows: Warm the solid until it melts. Weigh 5.81 g in a large beaker and dilute to 100 mL with 0.5 M NaOH solution (20 g of NaOH per liter). Wear gloves; hexamethylenediamine is absorbed through the skin.
2. Solution B: Adipoyl chloride, 0.25 M: Weigh 4.58 g and dilute to 100 mL with cyclohexane.

Teaching Tips

NOTES

1. The nylon can be colored by adding 1 drop of methyl red or bromcresol blue to the solutions.
2. If students want to keep the nylon, rinse it several times with water until it is free of all traces of amine.
3. Another nylon, Nylon 610, can be prepared as follows. (Be careful: Wear gloves; sebacoyl chloride is toxic and absorbed through the skin.)
   a. Solution A: 1 mL of sebacoyl chloride in 50 mL of trichlorotrifluoroethane.
   b. Solution B: 2.5 mL of hexamethylene in 25 mL of water.

c. Mix two volumes of A and one volume of B.

d. The reaction for producing Nylon 610 is

$$H_2N(CH_2)_6NH_2 + Cl\overset{O}{\overset{||}{C}}(CH_2)_8\overset{O}{\overset{||}{C}}Cl \xrightarrow{\text{NaOH}} H[NH-(CH_2)_6NH\overset{O}{\overset{||}{C}}-(CH_2)_8\overset{O}{\overset{||}{C}}]_xCl$$

Nylon 610

QUESTIONS FOR STUDENTS

1. Why is this called Nylon 66?

2. Why is the nylon synthesized at the interface of the two liquids?

3. What is a polymer?

102.   Synthesis of Rayon

Cellulose in filter paper is dissolved and, upon acidification, precipitated and regenerated as rayon.

Procedure

Wear safety goggles. This demonstration must be done in a hood.

1. Prepare this solution of copper(II) hydroxide prior to the demonstration:
   a. Place about 125 mL of water in a 500-mL beaker.
   b. Add copper sulfate and stir until the solution is saturated.
   c. With constant stirring, add concentrated aqueous ammonia drop by drop until a blue-green color appears and a precipitate forms. Do not add too much aqueous ammonia. If the solution becomes dark blue, start again.
2. With 11-cm filter paper, filter the solution you have prepared to recover the copper(II) hydroxide. Wash it once and discard the filtrate. It can be flushed down the sink.
3. Place the filter paper and the filtrate in a second 500-mL beaker. Shred two more pieces of filter paper and place these pieces in the beaker also.
4. Place the beaker on a magnetic stirrer and add 100 mL of concentrated aqueous ammonia. Continue stirring until the paper is all dissolved and the solution becomes thin enough to pour easily. This should take about 30 min. You may need to add a little more aqueous ammonia.
5. Place 200 mL of 0.5 M sulfuric acid in a shallow glass dish.
6. Carefully fill a pipet, syringe (with needle removed), or dropper with the blue solution.
7. Place the tip of the pipet, syringe, or dropper beneath the surface of the acid in the shallow dish and gently squirt the blue solution into the acid. If you use an even pressure, you will produce a fine thread of rayon.
8. Observe the rayon thread for a few minutes. After it turns white, remove the thread, wash it several times, and dry it.

Reactions

Copper hydroxide is prepared by reacting copper(II) sulfate with aqueous ammonia:

$$Cu^{2+} \text{ (aq)} + 2OH^- \longrightarrow Cu(OH)_2 \text{ (s)}$$
$$\text{blue-green}$$

The copper hydroxide reacts with additional aqueous ammonia to form tetraamminecopper(II) hydroxide, $Cu(NH_3)_4(OH)_2$. The filter paper (cellulose) dissolves in this complex and is regenerated upon acidification as the polymer rayon. Rayon is insoluble in aqueous solution, so it precipitates when it is squirted in the aqueous acid solution.

Solutions

1. Aqueous ammonia is concentrated. Be careful with this solution. Wear gloves, safety goggles, and a face shield when you use concentrated ammonia.
2. Sulfuric acid is 0.5 M. See Appendix 2 for directions.

Teaching Tips

NOTES

1. This process for making rayon is sometimes called the cuprammonium process.
2. Wood chips are the commercial source of cellulose.
3. Other, better ways are used to make rayon. In the viscose process, cellulose is treated with sodium hydroxide and then carbon disulfide. The cellulose is thus converted to viscose, a viscous yellow fluid. After aging, the viscose is squirted into dilute sulfuric acid and hydrogen sulfate solutions, where it forms rayon. In the cellulose acetate method, cellulose is treated with glacial acetic acid and acetic anhydride. The product is hydrolyzed and then dissolved in acetone. This viscous solution is forced through a spinneret into warm air to form cellulose acetate rayon.

QUESTIONS FOR STUDENTS

1. What is the appearance of the rayon?
2. What dissolves cellulose?
3. Could you use sources of cellulose other than filter paper?

103.   Synthetic Rubber

Two solutions are heated. When cooled, a rubbery mass precipitates. When squeezed and formed into a ball, this material bounces and shows other properties of rubber.

Procedure

Wear safety goggles and disposable gloves. Do this demonstration in a hood.

1. Place dark brown sodium polysulfide solution in a large beaker.
2. Heat the solution to 65–70 °C and stir in 1.0 g of $Mg(OH)_2$. Continue stirring until the magnesium hydroxide dissolves.
3. Slowly add 25–30 mL of ethylene chloride 1 mL at a time. You will notice an immediate reaction and the evolution of heat. Watch the temperature; do not let it exceed 80 °C. Stir this solution until the color changes from dark brown to a cloudy light brown. Continue to stir for about 15 min.
4. Remove the beaker from the heat and allow it to cool by standing at room temperature.
5. The polysulfide rubber should settle to the bottom of the beaker. Pour off the liquid and rinse the rubber several times with water.
6. The material should coagulate into one large mass. If it does not, add approximately 10 drops of 1 M HCl, stir, and rinse again.
7. Remove the rubber, squeeze it to remove water, form it into a ball, and bounce it on the laboratory bench.
8. Dispose of the waste according to the directions given in Appendix 5.

Reactions

1. NaOH and sulfur, S_8, react to form sodium polysulfide:

$$2Na^+ \text{ (aq)} + \tfrac{1}{2}S_8 \text{ (s)} \longrightarrow \underset{\substack{|| \; || \\ }}{\overset{\substack{S \; S \\ || \; ||}}{NaS\text{–}SNa}} \text{ (s)}$$

2. The reaction of sodium polysulfide with ethylene chloride produces a simple condensation polymer and sodium chloride as a byproduct:

$$Cl\text{-}CH_2\text{-}CH_2\text{-}Cl + Na\text{-}\overset{\substack{S\text{-}S \\ || \; ||}}{S\text{-}S}\text{-}Na \longrightarrow \left[CH_2CH_2S\text{-}\overset{\substack{S\text{-}S \\ || \; ||}}{S}CH_2CH_2SS \right]_n + 2NaCl$$

$$\text{polysulfide rubber}$$

Solution

Sodium polysulfide:

1. Add 10 g of NaOH to 150 mL of water in a beaker and boil it until the NaOH dissolves (it does not take long). Wear gloves and a face shield when you use concentrated NaOH.

2. Slowly add, constantly stirring, 20 g of sulfur.
3. Continue stirring until the solution changes from light yellow to dark brown. This change should take about 15 min.
4. Allow the solution to settle and cool. Pour off the dark brown sodium polysulfide.

Teaching Tips

NOTES

1. This reaction is of historic interest because it was the first synthetic rubber made in the United States.
2. The material produced in this demonstration is the starting material for the vulcanized rubber process.
3. Notice that a polymer consisting of repeating units of ethane and polysulfide is produced.

QUESTIONS FOR STUDENTS

1. Why is this called polysulfide rubber?
2. How are the properties of this product similar to those of ordinary rubber? How are they different?
3. What conditions seem to be important in this reaction?
4. Write the equation for this reaction.

104.   The Embedded Penny: Making Glyptal Resin Plastic

Phthalic anhydride and glycerol are heated in a test tube. The clear plastic that is formed is poured into an aluminum cap from a plastic cola bottle. A penny is placed in the cap and permanently embedded in the plastic.

Procedure

Wear safety goggles and disposable gloves. Do this demonstration in a hood.

1. Place 10 g of phthalic anhydride in a large test tube. Use an old test tube because it will be discarded at the end of the demonstration. Avoid breathing the vapors and avoid skin contact.
2. Add 0.5 g of sodium acetate to the test tube.
3. Add 4.0 mL of glycerol.
4. Heat the test tube slowly to dissolve any remaining solids.
5. Continue heating until the mixture boils for about 5 min.
6. Prepare a mold by cleaning and drying the aluminum cap from a large plastic cola container (the 3-L size works best).
7. Pour about half of the hot liquid into the aluminum cap.
8. Carefully place a penny (or some other object) in the liquid in the cap and then add the remaining liquid on top of the penny.
9. Let the plastic cool for about 10 min at room temperature. Then place it on top of a container of crushed ice for further cooling.
10. Using pliers, strip away the aluminum cap to free the plastic with its embedded penny.

Reactions

This is essentially an esterification reaction between the glycerol and phthalic anhydride. Water is produced during esterification and is removed by boiling the mixture. This plastic, glyptal resin, is a thermosetting plastic formed by phthalic anhydride acting as a cross-linker to hold strands of glycerol molecules together in a polymer.

Teaching Tips

NOTES

1. If all of the water formed during the reaction is not removed by boiling, the polymer will not be clear and may be sticky.
2. The final cooling of the plastic can be done in a refrigerator. It is important that it not become wet during cooling.
3. Other containers can be used as a mold. The shape of this plastic is determined by the mold. Polystyrene coffee cups or aluminum cupcake tins can be used.
4. During the esterification reaction, you will see bubbles produced by the formation of water.
5. Not all polymers are plastics. Some polymers are biomolecules such as carbohydrates and proteins.

6. Plastics, or more precisely, thermoplastic polymers, soften when heated to resolidify when cooled. The first commercially available plastic was Bakelite, a formaldehyde–phenol plastic discovered in 1909.
7. *Glyptal resin* is a thermoset polymer: It is permanently rigid and, once formed, will not melt when heated.
8. *Glyptal* is the term given to the resin formed between glycerol and phthalic acid. The term *resin* is used for a polymer that can be molded.
9. More than 300 thousand tons of phthalic anhydride (more than half of the total amount produced) is used to make polymers.
10. When the reaction appears to boil, water is actually boiling off.

QUESTIONS FOR STUDENTS

1. When the reaction appears to boil, what substance is actually boiling off?
2. How does the structure of the phthalic anhydride molecule change in the esterification reaction?
3. How does the structure of glycerol change?

Appendixes

Appendix 1. Periodic Chart of the Elements

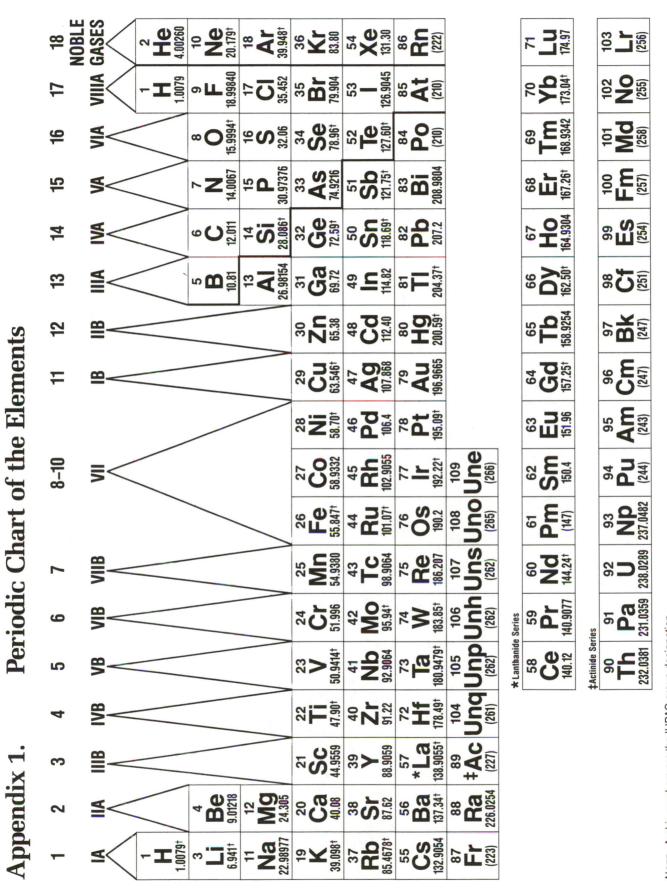

NOTE: Arabic numbers are the IUPAC group designation.
†Atomic weight varies because of natural variation in isotopic composition.

Appendix 2.

Properties and Preparation of Laboratory Acids and Bases

Parameter	Ammonium Hydroxide (NH_4OH)	Acetic Acid ($HC_2H_3O_2$)	Hydrochloric Acid (HCl)	Nitric Acid (HNO_3)	Sulfuric Acid (H_2SO_4)
Dilute this volume (in milliliters) of concentrated reagent to 1 L to make a 1.0 M solution	67.5	57.5	83.0	64.0	56.0
Dilute this volume (in milliliters) of concentrated reagent to 1 L to make a 3.0 M solution	200	172	249	183	168
Dilute this volume (in milliliters) of concentrated reagent to 1 L to make a 6 M solution	405	345	496	382	336
Normality of concentrated reagent	14.8	17.4	12.1	15.7	36.0
Molecular weight	35.05	60.05	36.46	63.02	98.08
Specific gravity	0.90	1.05	1.19	1.42	1.84
Approximate percentage in concentrated reagent	57.6	99.5	37.0	69.5	96.0

Note: To make *normal* solutions, use the same amount of reagent shown. However, to make a normal solution of sulfuric acid, use half the amount of reagent indicated. Example: dilute 28.0 mL of concentrated sulfuric acid to make 1 L of 1.0 N sulfuric acid solution.

Safety note: Concentrated acids and bases are corrosive. When you work with 6 M concentrations of acids or bases, wear gloves and safety goggles; work at a safety station equipped with a face shield or free-standing shield. When you dilute acids, always add the acid to water. Caution: Solutions will become hot. Use a fume hood when diluting or using concentrated acids or bases.

Appendix 3.   Equipment and Reagent List

This list includes the reagents and the equipment that are needed for each demonstration. Balances for weighing have not been listed. This list is for your convenience. You should always look at the individual demonstration and determine the concentration that works best for you on the basis of the purity and age of your stock reagents.

The numbers in italics indicate the number of the demonstration.

Equipment

Aluminum bottle cap   *104*
Aluminum foil   *22*
Aspirator with heavy rubber tubing   *13, 52*

Balance   *10, 11, 94, 95, 104*
Balloons   *3, 4, 9, 16, 55*
Bowl, large clear   *23*
Beaker, 50 mL   *40, 69, 75*
Beaker, 100 mL   *38, 47, 48, 62, 70, 84, 88, 100, 101*
Beaker, 250 mL   *5, 17, 18, 19, 27, 29, 32, 39, 41, 42, 47, 49, 54, 59, 62, 63, 64, 65, 71, 74, 80, 85, 97, 98*
Beaker, 600 mL   *28, 29, 33, 34, 51, 53, 57, 61, 67, 68, 81, 92, 102, 103*
Board, wooden   *1, 40*
Bottle, heavy plastic   *15*
Brush, artist's   *31*
Buret   *85*
Butane pocket lighter   *11*

Can, aluminum cola   *91*
Candle   *2*
Cardboard   *30, 96*
Clamp, buret   *12*
Cloth   *7*
Cotton   *8*
Crucible tongs   *82*
Cups, ceramic   *3*

Deflagrating spoon   *26*
Dialysis tubing   *93*
Dish, shallow glass   *102*
Dropper   *9, 25, 34, 57, 62, 73, 77, 78*

Evaporating dish   *73, 77, 82, 95*

File, triangular   *91*
Filter paper   *43, 73, 87, 95, 102*
Flashlight bulb with holder   *93*
Flask, Erlenmeyer, 50 mL   *75*
Flask, Erlenmeyer, 100 mL   *22*
Flask, Erlenmeyer, 250 mL   *16, 21, 43, 49, 66, 90, 95*
Flask, Erlenmeyer, 500 mL   *4, 55, 88*
Flask, filtering, 250 mL, with stopper   *6, 7, 13, 50, 52*
Flask, Florence, 500 mL   *5, 32, 36, 76*

Flask, Florence, 1000 mL   *84*
Floodlight   *86*
Funnel, small   *2*

Graduated cylinders, 10 mL   *69, 75, 87, 104*
Graduated cylinders, 100 mL   *8, 26, 35, 38, 42, 47, 48, 49, 54, 56, 60, 61, 62, 63, 64, 66, 67, 68, 74, 80, 92, 95, 97, 98, 99, 102, 103*
Graduated cylinder, 500 mL   *11, 56, 60, 99*

Hair dryer   *31, 45*
Hot plate   *24, 61, 82, 95*

Jar, large, with lid   *44*
Jar, large baby food   *93*

Leads, insulated wire, with alligator clips   *93*
Lecture bottle   *4, 16*
Litmus paper   *8*

Magnetic stirrer   *67, 68, 102*
Marshmallows   *13*
Mortar and pestle   *72, 94*

Nails   *15*
Napkin, heavy white   *79*

Paper, bond   *86*
Petri dish   *24, 46, 69, 83, 89, 96*
Plastic milk jug   *16*
Plastic wrap   *8*
Projector, overhead   *46, 89, 96*

Radio, 9-V   *93*
Ring stand and clamp   *5, 15, 45, 72*
Rubber bands   *45*

Self-sealing plastic bag   *10*
Steel wool   *6*
Stirring rod, glass   *70, 96*
Syringe, plastic, 50 mL   *12, 102*

Tape, cloth   *13*
Tesla coil   *15*
Test tubes   *6, 25, 37, 95*
Test tubes, 25mm $\times$ 200 mm, with stoppers to fit   *14, 20, 50, 51, 72, 87*
Thermometer   *39, 41, 42, 63, 103*
Toothpicks, round   *1, 58*

Reagents

Note: None of the reagents listed below is on the National Toxicology Program list of known or suspected carcinogens.

Appendix 4.   Safe Use of Chemicals

Most chemicals (and all chemicals in this volume) are safe to use **with reasonable care and in the amounts specified.**

A few substances fall into special care categories. It is a good idea to have only small amounts of these substances available to students and to keep supplies of larger amounts in a secure place to avoid mishap.

Irritants

Most chemicals can be irritating. Avoid inhaling dusts and fumes of chemicals and avoid splashing solutions on your skin or handling solids with your fingers. Work in the fume hood when you use chlorine, nitrogen dioxide, iodine, ammonia, sodium sulfide, trichlorotrifluoroethane, and other volatile substances. Other irritants include the metal chlorides of iron, aluminum, and copper; potassium permanganate; copper sulfate; calcium carbide; catechol; and 30% hydrogen peroxide.

Corrosives

Concentrated acids and bases are corrosive. Corrosive substances react with the skin. For demonstrations calling for 6 M or greater concentration of acids or bases, wear gloves and a face shield, as well as goggles. When you dilute solutions, always add the more concentrated solution to water. Use a fume hood when you dilute or use concentrated acids and bases.

Toxic Substances

Some substances should be used in small amounts because they are designated "toxic". Such substances include metal salts of barium, lead, silver, mercury, cobalt, and manganese and chlorine and mercury metal. The Federal Hazardous Substances Act has designated barium hydroxide, mercury metal, chlorine, and mercury(II) salts "highly toxic". Wear plastic disposable gloves when you mix solutions of these compounds. Whenever possible, it is wise to mix appropriate concentrations of stock solutions of these salts. Prepare stock solutions in the fume hood. Wear gloves and a dust mask when you handle the solids.

Some substances are no longer recommended for use with introductory chemistry programs. Laboratory and epidemiological studies show that evidence exists for the carcinogenicity of some chromium and nickel salts, for example. These substances are listed in the *Fourth Annual Report on Carcinogens* prepared by the National Toxicology Program (NTP) (U.S. Public Health Service, 1985). Although short-term exposure, as in an accidental spill, may pose little risk, we have chosen to avoid demonstrations using any substance on the NTP list.

See Appendix 5 for proper disposal procedures that prevent toxic substances from entering the environment.

Flammable Substances

When you use flammable liquids, extinguish all flames in the laboratory. We recommend a spark-proof hot plate for heating flammable liquids. Use the smallest possible amounts of flammable liquids.

Material Safety Data Sheets are provided with orders of chemicals. They contain detailed information about the use, toxicity, protective gear, and spill cleanup methods recommended for each chemical. Save them and keep them close to the laboratory. As you review these sheets, make them available to interested students in a designated area, not locked up.

Learn about American National Standards Institute (ANSI) standards and published recommendations for emergency and safety equipment, such as fume hoods, eyewash fountains, fire extinguishers, and safety goggles. Have the equipment inspected regularly.

The American Chemical Society maintains a referral service for people who want more information about health and safety. The Health and Safety Referral Service can be reached through

Maureen Matkovich
American Chemical Society
1155 16th Street, NW
Washington, DC 20036
202-872-4515

Other sources of information, including the booklet *Safety in Academic Chemistry Laboratories*, are available from the ACS Government Relations and Science Policy Department. Local colleges and universities usually have extensive libraries of books and journals describing accepted practices in use, storage, and disposal of chemicals.

Note: The designations "irritant", "corrosive", and "toxic substance" come from "School Science Laboratories, a Guide to Some Hazardous Substances", available from the U.S. Consumer Products Safety Commission, Washington, DC 20207.

Appendix 5. Chemical Disposal and Spill Guidelines

Disposal

Metal Compounds

The most environmentally responsible method of disposing of heavy metal compounds is to precipitate them and dispose of them in a managed hazardous waste disposal facility. Your state department of education or local university may have a program through which school laboratory waste materials are collected and managed. Use the following precipitation steps and store the precipitates until you can take them to an approved state or municipal landfill.

Form precipitates of the following metals in a beaker, then filter them and allow them to dry on the filter. Place the dry precipitate and filter paper in a covered jar. Pour the filtrate (liquid) down the drain with running water.

1. **Barium compounds.** Convert barium compounds to barium sulfate compounds. To prepare barium sulfate, add 10% sodium sulfate to a solution of barium salt. This precipitate is very fine. Although soluble barium compounds are toxic, barium sulfate is not considered a hazardous waste.

2. **Lead, cobalt, and silver compounds.** Convert these compounds to metal sulfide compounds. To prepare metal sulfides, add 10% sodium sulfide slowly until precipitation is complete.

3. **Solutions of other metal salts mentioned in this book.** Prepare dilute solutions and pour them slowly down the sink with running water.

Note: Many metal salts not called for in this book require special use and disposal techniques. These include mercury, arsenic, cadmium, chromium, and nickel.

Acids and Bases

Dilute small amounts of acids and bases by adding the acid or base to a large amount of water and pouring the liquid down the drain. Another method is to neutralize acid with dilute aqueous ammonia or baking soda solution. Neutralize bases with dilute hydrochloric acid or vinegar solution.

Organic Compounds

1. **Alcohols, organic acids, and acetone.** Dilute to 10 times their volume with water and then pour them slowly down the drain with running water.

2. **Waste hydrocarbon solvents and esters.** Do not place them in the sink. Consult your school maintenance supervisor for local solvent disposal procedures.

Metals and Nonmetals

Wash and store these substances for reuse. Do not place powdered metals or iodine in the solid waste. Some finely divided metals may react with damp paper to start a fire. Keep used mercury metal in a sealed jar for reuse.

These disposal guidelines comply with federal regulations. However, your local and state regulations may be different. Obtain information from your state department of environmental quality or from the Environmental Protection Agency (800-231-3075).

Accidental Spills of Irritants, Corrosives, Toxic Substances, or Flammables

Wear gloves and safety goggles. For spills in general, confine the spill, neutralize it, and mop it up. Confine the spill with cat litter (bentonite). Sand is a less effective alternative.

If an acid spills, neutralize the spill with sodium bicarbonate (baking soda), sodium carbonate, or calcium carbonate. Sodium bicarbonate or boric acid can be used to neutralize alkali spills.

Spilled mercury from broken thermometers should be collected and saved in a closed, labeled container. Special mercury spill kits are available from science supply houses. They offer the best assurance of picking up spilled mercury in cracks. Alternatively, dust powdered sulfur onto the spilled droplets and collect all of the material into a tightly capped jar, to be disposed of as mercury sulfide.

Appendix 6.   Guidelines for Safe Chemical Demonstrations

The Executive Committee of the ACS Division of Chemical Education has approved the following guidelines for chemical demonstrations performed at its meetings.

Interim Minimum Guidelines

Chemical demonstrators must

1. be familiar with the properties of the chemicals and with the chemical reactions involved in all demonstrations performed.

2. comply with all local rules and regulations.

3. wear some form of eye protection for all chemical demonstrations.

4. warn the members of the audience to cover their ears whenever a loud noise is anticipated.

5. plan the demonstration so that harmful quantities of noxious gases (e.g., nitrogen dioxide, sulfur dioxide, and hydrogen sulfide) do not enter the local air supply.

6. provide safety shield protection for any explosion whenever there is the slightest possibility that a container, its fragments, or its contents could be propelled with sufficient force to cause personal injury.

7. arrange to have a fire extinguisher at hand whenever the slightest possibility for fire exists.

8. NOT taste or encourage spectators to taste any non-food substance.

9. NOT use demonstrations in which parts of the human body are placed in danger (such as placing dry ice into the mouth or dipping hands into liquid nitrogen).

10. NOT use open containers of volatile, toxic substances (e.g., benzene, carbon tetrachloride, carbon disulfide, formaldehyde) without adequate ventilation as provided by fume hoods.

11. provide written procedure, hazard, and disposal information for each demonstration whenever the audience is encouraged to repeat the demonstration.

12. arrange for appropriate waste containers for and subsequent disposal of materials harmful to the environment.

Index

Index

Production and indexing: Paula M. Bérard
Book cover design: Carla L. Clemens
Managing editor: Janet S. Dodd

Typesetting: Hot Type Ltd., Washington, DC
Typeface: Melior
Printing: Maple Press Company, York, PA
Binding: Nicholstone Book Bindery, Nashville, TN

Recent ACS Books

Biotechnology and Materials Science: Chemistry for the Future
Edited by Mary L. Good
160 pp; clothbound; ISBN 0-8412-1472-7

Practical Statistics for the Physical Sciences
By Larry L. Havlicek
ACS Professional Reference Book; 512 pp; clothbound; ISBN 0-8412-1453-0

The Basics of Technical Communicating
By B. Edward Cain
ACS Professional Reference Book; 198 pp; clothbound; ISBN 0-8412-1451-4

The ACS Style Guide: A Manual for Authors and Editors
Edited by Janet S. Dodd
264 pp; clothbound; ISBN 0-8412-0917-0

Personal Computers for Scientists: A Byte at a Time
By Glenn I. Ouchi
276 pp; clothbound; ISBN 0-8412-1000-4

Writing the Laboratory Notebook
By Howard M. Kanare
146 pp; clothbound; ISBN 0-8412-0906-5

Principles of Environmental Sampling
Edited by Lawrence H. Keith
458 pp; clothbound; ISBN 0-8412-1173-6

Phosphorus Chemistry in Everyday Living, Second Edition
By Arthur D. F. Toy and Edward N. Walsh
362 pp; clothbound; ISBN 0-8412-1002-0

Chemistry and Crime: From Sherlock Holmes to Today's Courtroom
Edited by Samuel M. Gerber
136 pp; clothbound; ISBN 0-8412-0784-4

Folk Medicine: The Art and the Science
Edited by Richard P. Steiner
224 pp; clothbound; ISBN 0-8412-0939-1

For further information and a free catalog of ACS books, contact:
American Chemical Society
Distribution Office, Department 225
1155 16th Street, NW, Washington, DC 20036
Telephone 800-227-5558